Interfacial Physical Chemistry of High-Temperature Melts

Interfacial Physical Chemistry of High-Temperature Melts

Authored by
Kusuhiro Mukai

Translation supervised by
Taishi Matsushita

CRC Press
Taylor & Francis Group
Boca Raton London New York

CRC Press is an imprint of the
Taylor & Francis Group, an **informa** business

CRC Press
Taylor & Francis Group
6000 Broken Sound Parkway NW, Suite 300
Boca Raton, FL 33487-2742

First issued in paperback 2021

Interfacial Physicochemistry of High-Temperature Melts.
Originally published in Japanese by AGNE Gijutsu Center, Inc., Tokyo, Japan
Copyright ©2007 by Kusuhiro Mukai
All rights reserved.

ISBN-13: 978-0-367-21032-8 (hbk)
ISBN-13: 978-1-03-209071-9 (pbk)

Library of Congress Cataloging-in-Publication Data

Names: Mukai, Kusuhiro, editor. | Matsushita, Taishi, editor.
Title: Interfacial physical chemistry of high-temperature melts / [edited by] Kusuhiro Mukai, Taishi Matsushita.
Description: Boca Raton : Taylor & Francis, a CRC title, part of the Taylor & Francis imprint, a member of the Taylor & Francis Group, the academic division of T&F Informa, plc, [2020] | Includes bibliographical references.
Identifiers: LCCN 2019019492 | ISBN 9780367210328 (hardback : acid-free paper)
Subjects: LCSH: Interfaces (Physical sciences) | Materials at high temperatures.
Classification: LCC QC173.4.I57 I5838 2020 | DDC 541/.36--dc23
LC record available at https://lccn.loc.gov/2019019492

Visit the Taylor & Francis Web site at
http://www.taylorandfrancis.com

and the CRC Press Web site at
http://www.crcpress.com

Visit the eResources at
http://www.crcpress.com/9780367210328

Contents

Preface to the Japanese Edition

In our daily life, knowingly or unknowingly, we often see phenomena where the presence of an interface plays a dominant role, i.e., interfacial phenomena. For example, (1) a needle with an oil coating floats on water, (2) a piece of wood with camphor applied to one end begins to move on the water surface by itself (a so-called "camphor boat"), and (3) tears of wine (see Section 2.3.3 for details). Phenomenon 1 is mainly caused by the surface tension of water and the low wettability between the needle and water. Phenomena 2 and 3 are induced by the *Marangoni effect* (see Sections 2.3.3 and 2.5.2 ii and 4.2). Not only are such interfacial phenomena considered entertaining but they have also been proven by recent studies; they are closely related to important technological subjects in the processing of high-temperature materials.

Such phenomena, dominated by the existence of the interface, show up in a so-called *"interface-evolved world,"* where the existence of the interface cannot be ignored (see Section 1.2). The world that is treated in nanotechnology, which has been attracting attention recently, can also be included in the "interface-evolved world."

Meanwhile, not limited to interfacial phenomena, it is probably a standard approach to thoroughly observe an event and scientifically describe it to deeply understand it and comprehend its nature. Moreover, such a scientific approach can be a steady step in controlling various phenomena, solve problems, or achieve technological developments and improvements related to various technological subjects.

In the above-mentioned "scientific description of interfacial phenomena," the term "science" represents the title of this book, "interfacial physical chemistry" (see Section 1.1). Therefore, when we deal with the interfacial phenomena or various phenomena in the "interface-evolved world," it is especially important to acquire an ability beforehand to deeply understand the interfacial physical chemistry and apply it.

In Chapter 2 of this book, the fundamentals of interfacial physical chemistry are described to guide the readers and help them obtain a deeper understanding. To the best of my knowledge, the fundamentals of interfacial physical chemistry, such as surface tension, are still not completely understood by many researchers and engineers in the materials science and engineering field. Thus, it can reasonably be said that the understanding of surface tension is unclear worldwide. For this reason, surface tension is described in detail in Chapter 2. For a sufficient application of the fundamentals, I consider it necessary to understand the important equations through the derivation process. Therefore, derivation processes are also described in this chapter to some extent.

Chapter 3 briefly introduces the interfacial properties of high-temperature melts, which is the subject of high-temperature materials processing. This chapter is compiled so as to help the readers thoroughly understand Chapter 2 and apply the knowledge to Chapter 4.

In Chapter 4, examples of the application to materials processing at high temperature are described, focusing on the recent research results obtained by the author and his co-workers. Due to space limitations, many important studies by other researchers were unfortunately excluded. There are some research results and descriptions provided by the author and co-workers introduced in this book whose validity is the subject of future judgments. However, I have mentioned them as problem presentations and appreciate your understanding and patience.

Finally, I would like to thank the following individuals: Assoc. Prof. Toshiyuki Kozuka (Faculty of Engineering, Kumamoto University); Prof. Yutaka Shiraishi (Institute of Mineral Dressing and Metallurgy, Tohoku University); Dr. Masafumi Zeze (Yawata R & D Lab., Nippon Steel Corp.); Assoc. Prof. Tomio Takasu (Faculty of Engineering, Kyushu Institute of Technology); Prof. Taketoshi Hibiya (Faculty of Systems Design, Tokyo Metropolitan University); and Dr. Taishi Matsushita (Department of Materials Science and Engineering, Royal Institute of Technology (KTH), Sweden) for their contribution to examining the contents of the manuscript, collecting references, and so on, and to Ms. Yukari Izumi; Ms. Yoko Oosue; Dr. Olga Verezub; Ms. Hiroko Tanaka; Ms. Yoko Tonooka; Mr. Takahiro Furuzono; and Yukiko (my wife) for typing the manuscript, drawing figures, and so forth. For publication, Mr. Akikazu Maesono and Ms. Hisako Mihori (AGNE Gijutsu Center Inc.) have put in extraordinary efforts. I would like to express my sincere gratitude to them.

Kusuhiro Mukai
November 2006

Preface to the English Edition

This book was first published in 2007 by AGNE Gijutsu Center with the following title: *Kouon-yuutai No Kaimen-butsurikagaku*. This book is its English translation.

For the last 10 years, I presented the research results described in this book at every opportunity, and it attracted many researchers. For example, the theory and experimental results from microgravity experiments on the movement of fine particles caused by the surface tension gradient attracted researchers involved in the nozzle clogging problem in the continuous casting process. The work on the mechanism of the local corrosion of refractories has a good reputation, and the *in situ* observation of the penetration behavior of molten slag and molten metal into porous refractories has also received much recognition.

Professor Kusuhiro Mukai—the author of this book and my supervisor when I was a Ph.D. student—asked me to write and publish a revised enlarged edition of the above-mentioned Japanese book in English. However, in May 2018, while we were preparing the English edition, Professor Mukai passed away, and I felt that it was not appropriate to revise the book without his supervision. Then, encouraged by the graduate students of Prof. Mukai's lab, we collaborated to publish this English edition as a lasting tribute to his work.

This book is basically translated from the Japanese edition, but some notes have been added by the translation supervisor. The † symbol means that related video clips are provided at https://www.crcpress.com/9780367210328.

Throughout the preparation, I have received the cooperation from the Japanese edition publisher, AGNE Gijutsu Center. Moreover, last but not least, I wish to acknowledge the contributions to the publication of this English edition by Prof. Mukai's former students who graduated from the laboratory during his more than 30-year career at the Kyushu Institute of Technology.

Taishi Matsushita
Editor and Translation Supervisor
April 2019

Authors

Kusuhiro Mukai was a Professor Emeritus at the Kyushu Institute of Technology, Japan and Northeastern University, China. He received his Ph.D. from Nagoya University (1968) and became an Associate Professor at the Kyushu Institute of Technology (1969). He was a guest professor at University of Toronto, Canada (1985) and Imperial College London, UK (2005). He was a Professor at the Kyushu Institute of Technology from 1986 to 2004. His research area is high-temperature physical chemistry.

Taishi Matsushita is an Associate Professor at the School of Engineering, Jönköping University, Sweden, since 2012. He received his Ph.D. from Kyushu Institute of Technology (2003) and became a Senior Researcher at the Royal Institute of Technology (KTH), Sweden in the same year. He was given the title Docent (corresponding to Associate Professor) from KTH in 2008. His research area is high-temperature physical chemistry.

1 Introduction

1.1 INTERFACIAL PHYSICAL CHEMISTRY

Although I have not carefully retraced when the term "interfacial physical chemistry," which is used in the title of this book, began to be used in scientific literature, I believe that I started using it myself in my recent review article.[1] This phrase refers to the academic discipline where the equilibrium and kinetics of an interface are investigated mainly from a macroscopic viewpoint, using interfacial chemistry, chemical thermodynamics, and transport phenomena. Therefore, "interfacial physical chemistry" is now an established science as long as it does not exceed the scope of its application. The readers should understand interfacial physical chemistry as it stands now in this book, but the field of kinetics has not yet been fully systematized. Thus, the systematic understanding of interfacial physical chemistry depends on its future progress.

If we can deal with various phenomena and problems in engineering from an interfacial physical chemistry viewpoint by deeply understanding it and acquiring the capability of its application, we will be able to see a world that has not been seen so far. Furthermore, it will become a useful tool for opening a way to elucidate its relation to technological subjects and solve the problems in engineering.

1.2 INTERFACE-EVOLVED WORLD

How is the "interface-evolved world" dominated by interfacial phenomena? This has been elaborated.

We tend to consider the world we are experiencing every day to be "like this" based on our own perception of height, weight, and time as benchmarks. However, what will happen in a world where these benchmarks are extremely different from those of ours? One such example was introduced by G. Gamow, a famous scientist known for the big bang theory, in the book *Mr. Tompkins in Wonderland*.[2] He presented the following example: If we ride a bicycle in a world where the velocity of light is 20 km/h, what will happen? According to the theory of relativity, the surrounding scenery we are observing will become shorter or the speed of time on our watch will reduce. He described such a phenomenon with much interest.

The "interface-evolved world" refers to a world where the ratio of surface to bulk, i.e., specific surface area, has become abnormally large. More specifically, the world is composed of extremely small particles, thin films, or thin wires. Thermodynamically, the interface is located at the phase boundary. Contrary to the homogeneous phase, there exists a certain interfacial tension. According to the thermodynamic description, it can be interpreted that the interface has an excess Helmholtz energy, which will be explained later (see Section 2.2.2.1). In terms of mechanics, it can also be interpreted that the interfacial tension is the force per unit length that acts isotopically along the interface. In other words, the interfacial tension has two aspects: the thermodynamic aspect, the excess of Helmholtz energy per unit area, and the mechanistic aspect, the force per unit length. When the specific surface area is small, the contribution of the interfacial tension is small compared to the free energy of the bulk phase or a force originating from gravity. Conversely, in the "interface-evolved world" with a large specific surface area, the contribution of interfacial tension cannot be ignored, and various interfacial phenomena show up corresponding to the situation of the system. Besides the interfacial phenomena described in the "Preface" of this book, many phenomena that cannot be understood by our common-sense in daily life actually occurs. For example, a liquid thin film of slag goes up against gravity and fine bubbles in aqueous water sink. In addition, even in a scientific field, the well-known Gibbs' phase rule equation cannot be applied to the "interface-evolved world." For example, melting point, solubility, and vapor pressure depend on the radius of curvature of the interface. Therefore, the radius of curvature must be newly introduced as an intensive property that is variable for the system. The Gibbs' phase rule equation was originally introduced without considering the contribution of the interface; therefore, it is natural that the equation cannot be applied to the "interface-evolved world."

1.3   RELATION TO ENGINEERING

Note that interfacial phenomena are deeply related to many other aspects of engineering. One such example is the Marangoni effect, which has recently attracted much attention, and its connection with engineering is now being actively investigated.

With the development of satellites, a gravity term as a variable can be expanded from a conventional gravitational field on the ground to microgravity (so-called zero gravity). On the ground, the density convection generated by gravity cannot be eliminated but disappears under microgravity conditions. Meanwhile, in the process of making crystals from a molten state, a temperature gradient always exists. Therefore, the Marangoni convection caused by a surface tension gradient induced by the temperature gradient cannot be avoided. In contrast, under the condition where the surface tension gradient can be arbitrarily controlled, the Marangoni convection can be observed by separating it from the density convection. Owing to the advent of such research environment, research on Marangoni convection itself using the microgravity condition or that related to materials processing has been extensively conducted in the past. Some such studies were reviewed by an author in 1985.[3] Other research topics related to Marangoni convection include lubrication,[4,5] the abnormal acceleration phenomenon of aqueous surfactant solution in front of an overflow weir,[6] the behavior of liquid crystal,[7] relations with the convection of a liquid thin film (so-called Benard–Marangoni convection),[8] and the concept of infrared-visible image converters.[9]

The author's group has conducted research focusing on the relation between the interfacial phenomena and materials processing engineering at high temperatures, more specifically the metal refining processes, corrosion of refractories, growth processes of silicon single crystals, and welding, aiming to elucidate the relations between them and their technical problems. As a result, it has been clarified that interfacial phenomena are deeply (or possibly) related to these fields.

In Chapter 4 of this book, the relation between the interfacial phenomena and materials processing at high temperatures is introduced. In the majority of modern high-temperature materials processing such as iron and steelmaking processes, the processes, which one can produce a large quantity of homogeneous material (solid) from the liquid state through the solidification process by economical ways, are adopted. In this field, a liquid at high temperature is conventionally called a high-temperature melt. Based on the material, it is called molten metal, molten slag, molten salt, etc. In this book, to describe the liquid state, the terms "liquid" and "liquid phase" are mainly used in Chapter 2, and "melt," "molten metal," etc. are used in Chapters 3 and 4.

REFERENCES

1. Kusuhiro Mukai: *Bull. Iron Steel Inst. Japan*, **5** (2000), 725.
2. George Gamow: *Mr. Tompkins in Wonderland*, Cambridge University Press (1940).
3. Kusuhiro Mukai: *Tetsu-to-Hagané*, **71** (1985), 1435.
4. A. A. Fote, L. M. Dormant, and S. Fuerstein: *Lub. Eng.*, **32** (1976), 542.
5. Fujio Hirano and Tadao Sakai: *J. Japan Soc. Lubr. Eng.*, **22** (1977), 490.
6. Nobuyuki Imaishi. *Kagaku Kogaku Ronbun.*, **8** (1982), 136.
7. F. M. Leslie: *J. Phys. D Appl. Phys.*, **9** (1976), 925.
8. P. Cerisier, J. Pantaloni, G. Finiels, and R. Amarlic: *Appl. Opt.*, **21** (1982), 2153.
9. J. C. Loulergue, P. Manneville, and Y. Pomeau: *J. Phys. D Appl. Phys.*, **14** (1981), 1967.

2 Fundamentals of Treating the Interface

2.1 INTERFACE

The term "surface" used as a common word refers to the interface between solid and gas or that between liquid and gas. In a heterogeneous system composed of multiple phases, there is a part where the intensive properties (thermodynamic properties, such as density, which are unrelated to the mass of the system) exhibit discontinuity. In terms of thermodynamics, this part is defined as the interface.*

If the influence of the external force field (gravitational field, electromagnetic field, etc.) is negligible, each phase in a heterogeneous system is defined as the homogeneous part of the material systems that is distinguished from the others by a clear physical boundary. In other words, it is defined as the part where the intensive properties are uniform. However, from a microscopic perspective, the interface has a thickness that is thicker than at least one atom or molecule. Therefore, we can consider that the intensive property continuously changes within the thickness of the interface. Although the molecular dynamical approach has recently progressed to express the interface, including the liquid phase, which will be discussed in detail in this chapter at the microscopic scale, there remain difficulties in this approach. Therefore, the treatment of the interface hereafter proceeds based on the above-mentioned thermodynamically defined macroscopic perspective.

2.2 THERMODYNAMIC TREATMENT OF THE INTERFACE

2.2.1 GIBBS' METHOD

As described in Section 2.1, there remain several difficulties in the microscopic description of the interface. Even now, the fundamentals of the treatment of the interface, including the liquid phase, rely on Gibbs' method, as described below.

Gibbs introduced the concept of a virtual surface, i.e., the dividing surface (S), which has no thickness, in the region where the intensive properties show discontinuity, as shown in Figure 2.1. He then tried to macroscopically describe the interface based on this concept.

Figure 2.1 shows the α and β phases and their interface (perpendicular to the xy plane and parallel to the y-axis). The figure also shows the distribution of concentration c_i (mol/m^3) of component i in the α phase, β phase, and the interface. Gibbs introduced the dividing surface S, parallel to the interface and without thickness, in this interface region.

---

* Hereafter, the term "interface" is essentially used unless the surface needs to be distinguished from the interface.

5

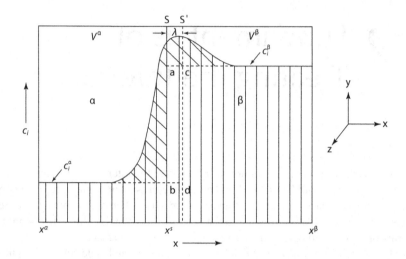

FIGURE 2.1 Gibbs' dividing surface.

As shown in Figure 2.1, let the total mol number of component i in $V^\alpha + V^\beta$ (total area multiplied by the thickness in Figure 2.1) be n_i (corresponding to the area of vertical line section multiplied by the thickness). If the concentration of component i in the α phase, c_i^α, does not change up to S, the mol number of component i is $c_i^\alpha V^\alpha$. Similarly, the mol number of component i in V^β is $c_i^\beta V^\beta$. Therefore, the value n_i^s, which is obtained by subtracting $(c_i^\alpha V^\alpha + c_i^\beta V^\beta)$ from the total mol number n_i, corresponds to the area of the shaded section in Figure 2.1, more precisely, to $\int_{x^\alpha}^{x^s} \left(c_i - c_i^\alpha \right) dx + \int_{x^s}^{x^\beta} \left(c_i - c_i^\beta \right) dx$, proving that the height and thickness of the frame in Figure 2.1 are of unit length (m). This can be considered as the mol number of the excess component i in the interface. Namely:

$$n_i^s = n_i - \left(c_i^\alpha V^\alpha + c_i^\beta V^\beta \right).$$ (2.1)

Assuming the area of the dividing surface S to be A, the excess amount of component i per unit interfacial area, i.e., the interfacial excess Γ_i (mol/m²), can be defined as:

$$n_i^s = A\Gamma_i.$$ (2.2)

Based on this method, the other interfacial amount (interfacial excess) can be defined in a similar manner.

Internal energy (corresponding to the surface energy, described later in Section 2.2.2.1) is given with:

$$U^s = U - \left(u_v^\alpha V^\alpha + u_v^\beta V^\beta \right).$$ (2.3)

Entropy is given with:

$$S^s = S - \left(s_v^\alpha V^\alpha + s_v^\beta V^\beta\right). \tag{2.4}$$

Helmholtz energy (corresponding to the surface tension, described later in Section 2.2.2.1) is given with:

$$F^s = F - \left(f_v^\alpha V^\alpha + f_v^\beta V^\beta\right). \tag{2.5}$$

Gibbs energy is given with:

$$G^s = G - \left(g_v^\alpha V^\alpha + g_v^\beta V^\beta\right), \tag{2.6}$$

where u_v^v, s_v^v, f_v^v, g_v^v ($v = \alpha, \beta$) are the internal energy, entropy, Helmholtz energy, and Gibbs energy, respectively, per unit volume of v phase.

In the Gibbs' method, various properties at the interface are not described from the microscopic perspective. Instead, it uses the quantities n_i^s, U^s, etc., which are the differences from those in the bulk phases α and β, for the properties at the interface, and integrate all of them in the dividing surface S. In this method, the value of n_i^s may change depending on the position of S, as shown in Figure 2.1, and such inconvenience is unavoidable. It is also inconvenient since one can only deal with surface excess.*

2.2.2 Surface Tension

2.2.2.1 Thermodynamic Interpretation of Surface Tension

Now, the first law of thermodynamics is applied to the surface of a unary liquid phase. When the surface of the system has increased by dA, the change in the surface internal energy dU^s is given with:

$$dU^s = dQ^s + \gamma dA \tag{2.7}$$

where dQ^s is the heat absorbed by the surface, γdA is the amount of isothermal work conducted at the surface due to change dA, and γ is the surface tension, which is described later.

For a reversible change (equilibrium):

$$dQ^s = TdS^s. \tag{2.8}$$

Therefore:

$$dU^s = TdS^s + \gamma dA. \tag{2.9}$$

---

* To resolve this inconvenience, Guggenheim introduced the concept of two dividing surfaces.[1] However, because this concept increased the complexity, Gibbs' method seems to be extensively used currently.

The Helmholtz energies F and F^s are expressed as:

$$F = U - TS,$$ (2.10)

$$F^s = U^s - TS^s.$$ (2.11)

Taking the total differentiation of the above equations, the following equation is obtained:

$$dF^s = dU^s - TdS^s - S^s dT$$

$$= TdS^s + \gamma dA - TdS^s - S^s dT$$ (2.12)

$$= \gamma dA - S^s dT.$$

Therefore:

$$f^s \equiv \left(\frac{\partial F^s}{\partial A}\right)_T = \gamma.$$ (2.13)

From Equation (2.13), it is understood that the surface tension is the surface excess of the Helmholtz energy per unit surface area.

According to the definition of the Helmholtz energy, the value of f^s is expressed as:

$$f^s = u^s - Ts^s.$$ (2.14)

In other words, the surface tension γ $(=f^s)$ is composed of the internal energy term u^s (surface energy) and entropy term (Ts^s), and u^s and s^s are the internal energy per unit surface area and surface excess of entropy, respectively.

Figure 2.2 shows the chemical bonding states of atoms–molecules in a liquid phase and at the surface, which are schematically represented on a two-dimensional plane. Since the number of chemical bonds representing the interaction among particles is smaller for molecules at the surface than that for molecules inside, the energy states of the surface molecules become higher according to the difference in the number of chemical bonds. This increase in the energy states corresponds to the value of u^s. As shown in Figure 2.2, existing states (ordered states) of molecules at the surface must be different from those inside because the chemical bonding states at the surface are different from those inside. This means that the entropy should also be different between the surface and the bulk. This difference corresponds to the value of s^s.

In the case of a multicomponent system, the relation between f^s and γ is expressed as:

$$f^s = \gamma + \sum_{i=1}^{r} \mu_i \Gamma_i,$$ (2.15)

where μ_i is the chemical potential of component i.

Equation (2.15) is derived as follows.

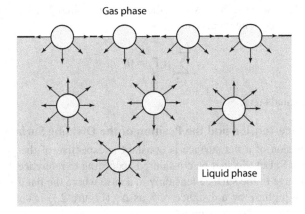

FIGURE 2.2 Schematic illustration of the bonding states of atoms or molecules inside the liquid phase and at the surface.

The total Helmholtz energy F of the system composed of the α phase (gas phase), β phase (liquid phase), and interface (area: A) at equilibrium is:

$$F^s = -p^\alpha V^\alpha - p^\beta V^\beta + \gamma A + \sum_{i=1}^{r} \mu_i \left(n_i^\alpha + n_i^\beta + n_i^s \right),$$

(2.16)

because at equilibrium:

$$\mu_i^\alpha = \mu_i^\beta = \mu_i^s.$$

(2.17)

In addition, when the variables in F except A are set as constant, Equation (2.13) can be used. This implies that the value γA is included in Equation (2.16) (for details, see references 2 and 3).

The Helmholtz energies of α phase, F^α, and β phase, F^β, are expressed as:

$$F^\alpha = -p^\alpha V^\alpha + \sum_{i=1}^{r} \mu_i n_i^\alpha,$$

(2.18)

$$F^\beta = -p^\beta V^\beta + \sum_{i=1}^{r} \mu_i n_i^\beta.$$

(2.19)

From Equation (2.5), the following is obtained:

$$F^s = F - F^\alpha - F^\beta = \gamma A + \sum_{i=1}^{r} \mu_i n_i^s.$$

(2.20)

By dividing both sides of the equation by A, Equation (2.15) is obtained.

If the dividing surface S is set at the position (called "equimolar surface") expressed by the following equation:

$$\sum_{i=1}^{r} \mu_i \Gamma_i = 0, \tag{2.21}$$

the value of f^s equals to γ.

2.2.2.2 Surface Tension and the Position of the Dividing Surface

The surface tension of a flat surface is constant irrespective of the position of the dividing surface S. Let this fact be considered according to reference 4.

Now, the change in the surface tension γ in a case where the flat dividing surface is moved to the β phase by a distance of λ as $\Delta\gamma$ (Figure 2.1) is considered. From Equation (2.15), the surface tension is:

$$\gamma = f^s - \sum_{i=1}^{r} \mu_i \Gamma_i = u^s - T s^s - \sum_{i=1}^{r} \mu_i \Gamma_i. \tag{2.22}$$

Meanwhile, the changes in u^s, s^s, and Γ_i in such a case are $\lambda(u_v^\beta - u_v^\alpha)$, $\lambda(s_v^\beta - s_v^\alpha)$, and $\lambda(c_i^\beta - c_i^\alpha)$, respectively. This can be easily understood considering the change in Γ_i in Figure 2.1. In other words, with the shift of λ, n_i^s corresponding to the area surrounded by abcd increases by Δn_i^s, which induces the change $\Delta n_i^s/A = \lambda(c_i^\beta - c_i^\alpha)/A = \Delta\Gamma_i$, where $u_v^\alpha = U^\alpha/V^\alpha$, $u_v^\beta = U^\beta/V^\beta$, $s_v^\alpha = S^\alpha/V^\alpha$, and $s_v^\beta = S^\beta/V^\beta$.

Therefore, from Equation (2.22), the change in the surface tension induced by the shift λ, $\Delta\gamma$ is:

$$\Delta\gamma = \lambda\left(u_v^\beta - u_v^\alpha\right) - T\lambda\left(s_v^\beta - s_v^\alpha\right) - \sum_{i=1}^{r} \mu_i \lambda\left(c_i^\beta - c_i^\alpha\right). \tag{2.23}$$

By rearranging the above equation, the following is obtained:

$$\Delta\gamma = \lambda\left\{u_v^\beta - T s_v^\beta - \sum_{i=1}^{r} \mu_i c_i^\beta - \left(u_v^\alpha - T s_v^\alpha - \sum_{i=1}^{r} \mu_i c_i^\alpha\right)\right\}. \tag{2.24}$$

Since $u_v^\beta - T s_v^\beta = f_v^\beta$, $\sum_{i=1}^{r} \mu_i c_i^\beta = g_v^\beta$,* $u_v^\alpha = T s_v^\alpha = f_v^\alpha$, $\sum_{i=1}^{r} \mu_i c_i^\alpha = g_v^\alpha$, and $f_v^\alpha - g_v^\alpha = -p^\alpha$ (which is the same for the β phase):

$$\Delta\gamma = \lambda\left\{f_v^\beta - g_v^\beta - \left(f_v^\alpha - g_v^\alpha\right)\right\} = \lambda\left(p^\alpha - p^\beta\right), \tag{2.25}$$

where $f_v^\alpha = F^\alpha/V^\alpha$ and $g_v^\alpha = G^\alpha/V^\alpha$ (which is the same for the β phase).

---

* This is because $G = \sum_{i=1}^{r} \mu_i n_i$ and $c_i^\beta = n_i^\beta/V^\beta$.

When the interface is flat, $p^\alpha = p^\beta$ at equilibrium. Therefore, the value of γ becomes constant irrespective of the position of the dividing surface S (the value of λ), i.e., $\Delta\gamma = 0$.

《In the Case of a Curved Surface》

Conversely, in the case of a curved surface (spherical surface), the surface tension changes depending on the position of the dividing surface S, as shown below:

$$\gamma = \gamma_s \left(\frac{r_s^2}{3r^2} + \frac{2r}{3r_s} \right), \tag{2.26}$$

where r is the radius of the dividing surface S. From Equation (2.26) and Figure 2.3, it is understood that the value of γ becomes minimum γ_s at $r = r_s$. The dividing surface S at $r = r_s$ is called "surface of tension."

Here note that the surface of tension is in the interfacial region. Therefore, as long as the thickness of the interfacial region (hereafter "interfacial thickness") δ_i (it may be approximately a few molecular layers or less) is smaller than r_s:

$$\gamma = \gamma_s \left\{ 1 + \varepsilon^2 + O(\varepsilon^3) \right\}. \tag{2.27}$$

where $\varepsilon = \delta_i / r_s$.

Therefore, the change in γ in the interfacial thickness δ_i is so small that γ can be assumed to be almost the same as γ_s.

For more details on the treatment of the curved surface, refer to reference 5 and reference 3, which explains an important part of the former easily.

2.2.2.3 Surface Tension and Radius of Curvature

In Section 2.2.2.2, it was described that surface tension depends on the position of the dividing surface S when the value of the curvature radius of the interface

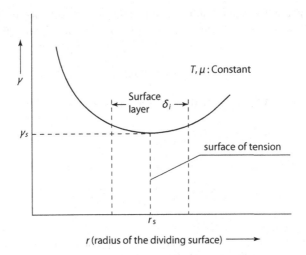

FIGURE 2.3 Changes in surface tension due to the position of the dividing surface.

becomes close to that of the interfacial thickness. Then, how does the surface tension change when the radius of curvature itself changes? Let the surface tension at the surface of tension be considered, which is practically interesting.*

As is discussed later, Laplace's Equation (2.28) (for a sphere) holds even when the radius of curvature is extremely small at the surface of the tension:

$$p^\alpha - p^\beta = \frac{2\gamma_s}{r_s},\tag{2.28}$$

where γ_s is the surface tension at radius r_s at the surface of tension. At equilibrium:

$$\mu^\alpha = \mu^\beta = \mu^s.\tag{2.29}$$

In this equation, the value of μ^s is not an excess amount but the chemical potential (absolute value) at the interface. Considering the reversible small change δ,[§] equations (2.28) and (2.29) are, respectively, converted to equations (2.30) and (2.31):

$$\delta p^\alpha - \delta p^\beta = \delta\left(\frac{2\gamma_s}{r_s}\right),\tag{2.30}$$

$$\delta\mu^\alpha = \delta\mu^\beta.\tag{2.31}$$

Meanwhile, as is well known, the Gibbs–Duhem equation for α and β phases is given with:

$$s^\alpha\delta T - v^\alpha\delta p^\alpha + \delta\mu^\alpha = 0$$

$$s^\beta\delta T - v^\beta\delta p^\beta + \delta\mu^\beta = 0,\tag{2.32}$$

where v^α and v^β are the molar volumes of the α and β phases, respectively. When the temperature T is constant, the following relation is obtained from equations (2.31) and (2.32):

$$v^\alpha\delta p^\alpha = v^\beta\delta p^\beta\tag{2.33}$$

From equations (2.33) and (2.30), the following are derived:

$$\delta\left(\frac{2\gamma_s}{r_s}\right) = \frac{v^\beta - v^\alpha}{v^\alpha}\delta p^\beta,\tag{2.34}$$

$$\delta\left(\frac{2\gamma_s}{r_s}\right) = \frac{v^\beta - v^\alpha}{v^\beta}\delta p^\alpha.\tag{2.35}$$

---

* Here, because the pure component system (for example, component i) is considered, the suffix i for μ, v, Γ, s, etc., are omitted. Hereafter, the same notation will be used.
§ Note that the same symbol δ will be used later for the meaning of the δ phase.

By rearranging the left side of Equation (2.35), the following is obtained:

$$\frac{2}{r_s}\delta\gamma_s + 2\gamma_s\delta\left(\frac{1}{r_s}\right) = \frac{v^\beta - v^\alpha}{v^\beta}\delta p^\alpha. \tag{2.36}$$

As is described later (see Equation (2.72)), the Gibbs–Duhem equation including the interface is given as:

$$\delta\gamma_s = -s^s\delta T - \Gamma\delta\mu. \tag{2.37}$$

When the temperature is constant:

$$\delta\gamma_s = -\Gamma\delta\mu, \tag{2.38}$$

where Γ is the excess of the pure component at the surface of tension. From the relation $(\partial\mu/\partial p)_T = v$, $\delta\mu = v^\alpha\delta p^\alpha$ is obtained. When this relation is substituted in Equation (2.38), the term δp in Equation (2.36) is eliminated. Furthermore, rearranging Equation (2.36) using the relations $c^\alpha = 1/v^\alpha$ and $c^\beta = 1/v^\beta$:

$$\frac{\delta\gamma_s}{r_s} = \frac{2\Gamma}{\dfrac{2\Gamma}{r_s} + c^\alpha - c^\beta}\delta\left(\frac{1}{r_s}\right). \tag{2.39}$$

When the β phase is a gas phase and the α phase is a liquid phase (non-compressed liquid), $c^\alpha \gg c^\beta$ and the value of $c^\alpha = 1/v^\alpha$ can be considered to be constant. Then, by integrating Equation (2.39), the following Equation (2.40) is obtained:

$$\frac{\gamma_s}{\gamma_\infty} = \frac{c^\alpha}{\dfrac{2\Gamma}{r_s} + c^\alpha} = \frac{1}{2\dfrac{\Gamma}{c^\alpha}\cdot\dfrac{1}{r_s} + 1}. \tag{2.40}$$

Let the distance between the surface of tension and equimolar surface be $\lambda_\circ$ (this is the case where $\lambda = \lambda_\circ$ in Figure 2.1). Then, the following is obtained:

$$\lambda_\circ\left(c^\alpha - c^\beta\right) = \lambda_\circ c^\alpha = \Gamma. \tag{2.41}$$

Rearranging Equation (2.40) using $\lambda_\circ$:

$$\frac{\gamma_s}{\gamma_\infty} = \frac{1}{\dfrac{2\lambda_\circ}{r_s} + 1}. \tag{2.42}$$

The above equation is known as the Tolman equation.

For water, $\lambda_\circ \fallingdotseq 0.1$ nm can be estimated. This means that the change becomes $\gamma_s/\gamma_\infty \fallingdotseq 0.98$ when $\lambda_\circ/r_s = 0.01$, i.e., $r_s = 10$ nm and $\gamma_s/\gamma_\infty \fallingdotseq 0.83$ when $\lambda_\circ/r_s = 0.1$, i.e., $r_s = 1$ nm. However, it is necessary to carefully consider as to whether the conventional macroscopic thermodynamics, e.g., classical nucleation theory, can be applied to droplets with a radius of 10 nm or less.

Based on the above results, it can be understood that the radius-dependency of surface tension must be considered when the critical nucleation radius at the initial stage of nucleation is in the order of nanometers, or when the size of the material is so small like the ones treated in contemporary nanotechnologies, such as droplets and bubbles.

2.2.2.4 Surface Tension and Binding Energy

As described in Section 2.2.2.1, the surface tension of a unary liquid is composed of the internal energy term u^s and the entropy term Ts^s. Therefore, it cannot be simply said that the surface tension is equal to the binding energy. Accordingly, when the relation between the surface tension and the binding energy is interpreted, the contribution of the entropy term must be considered. Estimation of the entropy term is generally more difficult than that of the internal energy term. The example below, although it is an approximation, can be helpful to avoid the contribution of the entropy term. From equations (2.13) and (2.14), the following relation is derived:

$$\gamma = u^s - Ts^s. \tag{2.43}$$

When $T \rightarrow 0$, $\gamma = u^s$ is obtained. For estimating the value of u^s, the enthalpy of vaporization ΔH_{vap} can be used, which is easy to measure, i.e.,

$$\Delta H_{vap} = \Delta U_{vap} + P \Delta V_{vap} = \Delta U_{vap} + \Delta n_{vap} RT, \tag{2.44}$$

where ΔU_{vap} is the vaporization (internal) energy, ΔV_{vap} is the volume of the evaporated component, Δn_{vap} is the mol number of the evaporated component, and R is the gas constant. When $T \rightarrow 0$, $\Delta H_{vap} = \Delta U_{vap}$. When the average atomic number of the nearest atoms in a liquid phase is denoted as Z and those at the surface as Z^s, the value of $(Z - Z^s)/Z \times \Delta h_{vap}$ is equal to the excess internal energy u^s at the surface, proving that $\Delta H_{vap}/\Delta n_{vap} = \Delta h_{vap}$.

When the surface area per mol of the pure component is A_{mol}, the surface tension γ_{mol} per A_{mol} is equal to $A_{mol} \gamma$. Thus:

$$\gamma_{mol} = \frac{Z - Z^s}{Z} \Delta h_{vap}. \tag{2.45}$$

When it is assumed that the structure of the liquid is face-centered cubic, $Z = 12$ and $Z^s = 9$. Thus:

$$\gamma_{mol} = \frac{1}{4} \Delta h_{vap}. \tag{2.46}$$

Figure 2.4[6] shows the plot of the relation between $\gamma_{i, mol,0}$ and $\Delta h_{i, vap,0}$ by extrapolating the measured values of various pure liquid metals $\gamma_{i,mol}$ and $\Delta h_{i,vap}$ to the limit $T \rightarrow 0$. As seen in the figure, although the slope of the relation between the two values slightly deviates to smaller side compared to that of Equation (2.46), i.e., the

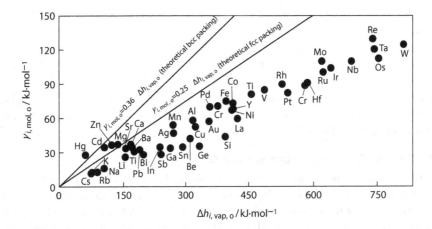

FIGURE 2.4 Relation between $\gamma_{i,\,mol,}$ and $\Delta h_{i,\,vap}$ at $T \to 0$. The suffix 0 represents the value when $T \to 0$. (Adapted from reference 6.)

line with slope 1/4, it can be said that the relation is close to the linear relation with slope 1/4.*

2.2.2.5 Surface Tension and Temperature

For the pure-component liquid, Equation (2.43) holds. Therefore, by partially differentiating both sides of the equation with respect to temperature T:

$$\left(\frac{\partial \gamma}{\partial T}\right) = \left(\frac{\partial f^s}{\partial T}\right) = -s^s. \tag{2.47}$$

Equation (2.47) shows that the temperature coefficient of the surface tension is equal to the value where the surface excess entropy s^s is multiplied by -1. Since the temperature coefficient of the surface tension in the pure-component liquid is generally a negative value, $s^s > 0$. Therefore, one can say that the entropy of the surface is higher than that of the bulk phase, i.e., the randomness is large.

Several empirical equations have been proposed for the relation between surface tension and temperature. Eötvös proposed the following Equation (2.48) in 1886:

$$\gamma = \frac{k}{v^{l^{2/3}}}(T_c - T), \tag{2.48}$$

where v^l is the volume of liquid per mol (molar volume), and T_c is the critical temperature. The value of k is about 2.1×10^{-7} J/deg for most liquids. Equation (2.48) has a disadvantage in that it does not fit near the critical temperatures.

---

* In Figure 2.4, apart from the errors in γ_{mol} and Δh_{vap} themselves, there are other errors caused by extrapolating these values to $T \to 0$. Therefore, the slope has an error that originates from these errors. Based on the discussion presented in Section 2.2.2.4, other approaches for obtaining the relation between surface tension and binding energy are possible and will be presented in Section 3.2.1.

The Katayama–Guggenheim Equation (2.49) is well known to have improved this disadvantage:

$$\gamma = \gamma_\circ \left(1 - \frac{T}{T_c}\right)^{11/9},\tag{2.49}$$

where $\gamma_\circ$ is a constant that varies depending on the type of liquid. Equation (2.49) agrees well with the experimental results for relatively simple liquids such as neon, argon, nitrogen, and oxygen.

2.2.2.6 Surface Tension and Surface Stress

In the previous sections, surface tension is discussed from the thermodynamic viewpoint. However, surface tension has a mechanical aspect, as the name implies, which is described later in Section 2.3. With respect to the surface tension as a force, the surface tension can be considered to be equal to the surface stress for liquid–gas, and liquid–liquid interfacial tension, as explained in the following section. However, in the case of the interface containing the solid phase, such as liquid–solid interface, the contribution of the stress must be considered. In this regard, let the relation between surface tension and surface stress be clarified.

Generally, two different quantities can be defined concerning the force or the work to expand a surface. One is a method to expand the surface without changing the distortion state of the surface. In other words, it is a method to expand the surface by forming a new surface whose state is the same as the one that already exists. In this case, the number of atoms at the surface increases in proportion to the surface area. As described in Section 2.2.2.1, the work δW required to reversibly expand the surface by δA at constant temperature and pressure is represented by $\delta W = \gamma \delta A$, where γ is the surface tension. This means that the surface tension γ corresponds to the work that is required to increase the surface area by a unit area.

The other method is to expand the surface by changing its surface distortion state, meaning that the surface is expanded by stretching the distance between atoms at the surface. In this case, the number of atoms at the surface does not increase. The reversible work δW, necessary to expand the surface in this method, is expressed by the following equation, where ζ is called the surface stress:

$$\delta W = \zeta \delta A.\tag{2.50}$$

The surface stress ζ is the mechanical resistance when the surface is elastically expanded and has a dimension of force per unit length.

For a liquid also, the surface stress can be defined as the force that is necessary to elastically expand the surface. In this case, the difference between surface tension and surface stress is eliminated due to the high mobility of molecules. When a liquid surface is stretched, the molecule density at the surface decreases. Thus, molecules are immediately transferred from the bulk, and the surface state returns to the initial state. Therefore, the surface tension γ is numerically equal to the surface stress ζ for the unary liquid system.

In the Laplace and Kelvin equations described later, the mechanical balance or changes in the chemical potential due to pressure are involved; therefore, strictly describing, the surface stress must be used when the interface containing a solid phase is treated.

Next, let the relation between surface tension and surface stress defined in the above discussion be quantitatively considered. When the number of atoms in a surface area A is denoted as N, the area "a" occupied by one atom is expressed as $a = A/N$. Then, if the surface free energy per atom is denoted as ϕ, the surface tension γ is represented as $\gamma = \phi/a$. Now, the case in which a homogeneous material with a flat surface is uniformly expanded at a constant temperature, pressure, and the total number of atoms is considered. Let the surface area be assumed to become $A + \delta A$ and the number of atoms on a surface to became $N + \delta N$. The work required for this procedure, δW, is used for expanding the surface overcoming the "effective" surface stress ζ' and stored at the surface. An increase in the surface area in this process is generally caused by an increase in both the number of surface atoms N and area "a" occupied by one atom. Therefore, the value of ζ' is not equal to that of γ or ζ but will lie in between the two. In this case, the work δW is expressed as:

$$\delta W = \zeta' \delta A = \delta(\phi, N) = \phi \frac{\partial N}{\partial A} \delta A + N \frac{\partial \phi}{\partial A} \delta A$$

$$\zeta' = \phi \frac{\partial N}{\partial A} + N \frac{\partial \phi}{\partial A}. \tag{2.51}$$

In the case of perfect plastic deformation due to the extremely high atomic mobility, such as in a liquid, the surface immediately returns to its initial state when expanded. In other words, the state of the atoms at the surface does not change because of the deformation. In this case, $(\partial \phi / \partial A)_a = 0$ and $(\partial N / \partial A)_a = 1/a$ when $a = \text{const}$. Thus:

$$\zeta' = \phi \left(\frac{\partial N}{\partial A} \right)_a = \frac{\phi}{a} = \gamma. \tag{2.52}$$

This equation shows that, in this case, the "effective" surface stress ζ' is equal to the γ value given by the N/m unit.

Now, let another extreme case be considered, i.e., the perfect elastic deformation. In this case, $N = \text{const.}$, which leads to $\partial N / \partial A = 0$. Thus, the first term on the right side of Equation (2.51) is eliminated, and $\delta A = N \delta a_0$, where a_0 is the area occupied by one atom in a stable state. Therefore:

$$\zeta' = \left(\frac{N \partial \phi}{\partial A} \right)_N = \left(\frac{N \partial \phi}{N \partial a} \right)_N = \left(\frac{\partial \phi}{\partial a} \right)_N = \zeta. \tag{2.53}$$

Since $\phi = \gamma a$, substituting it into the above equation, the following is obtained:

$$\zeta' = \left(\frac{\partial (\gamma a)}{\partial A} \right)_N = \gamma + \left(\frac{a \partial \gamma}{\partial a} \right)_N = \gamma + \left(\frac{\partial \gamma}{\partial \varepsilon} \right)_N = \zeta, \tag{2.54}$$

where $\partial\varepsilon(=\partial a/a)$ is the elastic distortion. The value of ζ' in this case is equal to the surface stress ζ described in the beginning, having a relation with surface tension γ, as expressed by Equation (2.54). Also, as can be seen from Equation (2.53), the surface stress ζ corresponds to the slope of the surface free energy against the area occupied by one atom (this can be considered as the distance between atoms).

Since $\phi=\gamma a$ and $\gamma=$const. for a liquid, $\partial\phi/\partial a=\gamma=\zeta$. However, in the case of a crystalline solid, the surface has an ordered atomic arrangement that is continuous from the inside. Thus, stress exists at the surface. For example, if only one atomic layer at the surface is separated from the solid and placed in a vacuum, the force from the inner atoms is not exerted, and the atomic area in the stable state $a_\circ$ is, therefore, different from that in a. When this atomic layer is returned to the solid surface, the compressive stress in the case of $a_\circ > a$ or the tensile stress in the case of $a_\circ < a$ must be applied to maintain the continuous atomic arrangement from the inside of the solid. Correspondingly, the unevenly distributed tensile or compressive stress is generated inside the crystal. If the compressive stress exists at the surface, the surface free energy ϕ decreases due to the increase in a, resulting in a stable state. Therefore, as can be seen from Equation (2.53), the value of ζ becomes negative. In contrast, if the tensile stress exists, the value of ζ becomes positive.

For more explanation of the topics given in Section 2.2.2.6, refer to references 7 and 8.

2.3 MECHANICAL TREATMENT OF INTERFACE

2.3.1 Mechanical Interpretation of Surface Tension

The model shown in Figure 2.5 is often used as an example to explain and understand the existence of surface tension and its mechanical meaning. As shown in the figure, a liquid film is formed on a U-shaped frame, and part of the film is supported by a fine wire AB that can move left and right without friction with the frame. According to the observation, if the wire is left, it moves spontaneously to the left. This wire stops only when it is pulled to the right with force $\mathbb{F}_\gamma$, i.e.,

$$\mathbb{F}_\gamma = 2\gamma^1 l, \tag{2.55}$$

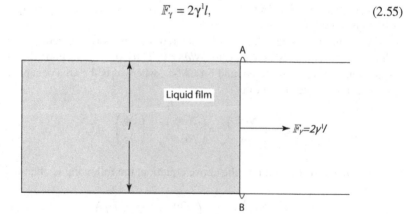

FIGURE 2.5 Force caused by the surface tension of the liquid film.

where l is the width of the U-shaped frame, and γ^l is the force acting on the surface per unit length, i.e., surface tension. In this case, the surface tension acts on both sides of the film. Therefore, factor 2 is multiplied on the right side of Equation (2.55). Since the surface of the liquid is isotropic, the surface tension of liquid γ^l can be regarded as an isotropic force acting on the surface per unit length.

Let the above-mentioned description of the mechanical perspective of surface tension be examined in more detail to define it. Generally, the pressure in a liquid is expressed by a tensor. In a static liquid (bulk phase), the pressure is isotropic and perpendicular to all planes considered. The value of the pressure is equal to p. When the pressure is represented by a symmetrical tensor, $p_{xx}, p_{yy}, p_{zz}, p_{xy}, p_{yz}$, and p_{zx}, these values can be described as $p_{xx}=p_{yy}=p_{zz}=p$ and $p_{xy}=p_{yz}=p_{zx}=0$.

Meanwhile, at the boundary between two phases, i.e., the interface, the above tensor becomes anisotropic due to the existence of surface tension.

If the z-axis is taken perpendicular to the flat interface,* the values of p_{xy}, p_{yz}, and p_{zx} become zero because of the symmetry of the interface. Additionally, $p_{xx}=p_{yy}$ and only the value of p_{zz} is different from those of the others. Since the values of p_{xx} and p_{yy} should be a function of z:

$$p_{xx} = p_{yy} = p_T(z), \quad p_{zz} = p_N(z) = p. \tag{2.56}$$

Since the liquid is assumed to be in its static state, i.e., equilibrium, the right side of Equation (2.56), i.e., $p_N(z)$, should be equal to the constant value of p irrespective of z. The value $p_N(z)$ represents the normal direction, i.e., perpendicular direction, of the interface. Hence, it is called normal pressure. Meanwhile, $p_T(z)$ is called tangent pressure, which acts in a plane that is parallel to the interface for a flat interface.

Based on the above preliminary discussion, let the plane that is parallel to the yz plane shown in Figure 2.6 be considered. The horizontal width of the plane has a unit length, and the vertical length is l, i.e., the vertical length from $l/2$ to $-l/2$. Herein, the length l is assumed to be longer than the interfacial thickness. The force $\mathbb{F}_{\gamma,x}$ acting in a direction vertical to this plane (i.e., the surface of this paper), i.e., parallel to the x-axis, is therefore expressed by the following equation:

$$\mathbb{F}_{\gamma,x} = -\int_{-l/2}^{l/2} p_T(z)dz. \tag{2.57}$$

Here, in Figure 2.6, the perpendicular direction of the front side of this paper is the positive side of the x-axis. If the surface tension does not exist, the force acting from the positive side of the x-axis is equal to $-pl$. Thus, the force generated by the existence of surface tension is the difference between Equation (2.57) and $-pl$, that is:

$$\gamma = -\int_{-l/2}^{l/2} p_T(z)dz - (-pl) = \int_{-l/2}^{l/2} \left(p - p_T(z)\right)dz. \tag{2.58}$$

---

* Note that for convenience, the x-axis is considered to be perpendicular to the interface in Figures 2.1 and 2.11 (shown later).

As can be understood from the above discussion, it is interpreted that γ exists within the interfacial thickness and the surface tension γ defined by Equation (2.58) acts on the line of unit length at the interface (horizontal width of Figure 2.6). Since the length l is set larger than the interfacial thickness, the result will be the same even if the length is larger than that, for example, even if the upper and lower limits of the integral are set to infinity. Then, Equation (2.58) can also be expressed as:

$$\gamma = \int_{-\infty}^{\infty} \left(p - p_T(z) \right) dz. \tag{2.59}$$

Since $p = p_N$ from Equation (2.56), the above Equation (2.59) can also be expressed as Equation (2.60):

$$\gamma = \int_{-\infty}^{\infty} \left(p_N - p_T(z) \right) dz. \tag{2.60}$$

Equation (2.60) is called Bakker's equation.

Owing to the improvement of the performance of computers, the field of molecular dynamics has progressed significantly in recent years. The molecular dynamics approach has been applied to a certain extent to describe the surface of the liquid. Therefore, if $p_T(z)$ can be more accurately described using molecular dynamics, the theoretical calculation of surface tension and the accompanying elucidation of the surface tension itself will progress from this front.

2.3.2 LAPLACE'S EQUATION

When the mechanical image of surface tension or liquid–liquid interfacial tension based on the description given in Section 2.3.1 is summarized, the mechanical image can be described as follows. In the macroscopic description, the surface is separated by a film having no thickness and the isotropic force per unit length acts on the

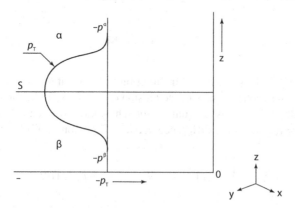

FIGURE 2.6 Changes in pressure on a flat plane parallel to a surface (tangential pressure) along the z-axis (vertical direction).

surface along the surface. Owing to this mechanical action of surface tension, various surface (interfacial) phenomena occur, as described below. The basic equation that describes such interfacial phenomena based on mechanical action is Laplace's equation.

To begin with, let the situation where a portion of bubbles is attached to a solid surface and mechanically balanced be considered, as shown in Figure 2.7. In this case, gravity can be ignored, and the surface of the bubble can be assumed to be a part of the sphere. Let the inside of the bubble assumed to be the α phase, the liquid phase to be the β phase (i.e., the surroundings of the α phase), and the pressure to be p^α and p^β, respectively. The force acting vertically on the δA portion of the surface is $(p^\alpha-p^\beta)\delta A$, and the z-axis component, i.e., Cz direction, of the force is $(p^\alpha-p)$ cos $\theta\cdot\delta A$. Through simple geometrical consideration, $\delta A\cos\theta=\delta A'$, where A' is the area of A projected onto the solid surface. Therefore, the z-component of the force acting on the entire surface area of this bubble (radius: r) is the sum of $(p^\alpha-p^\beta)\delta A'$. According to Pascal's law, the p^α and p^β values are constant. Thus, $\delta A'$ only has to be summed up. As a result, the total sum of $\delta A'$ amounts to $A'=\pi d^2$. Meanwhile, at the boundary where the solid and bubble are in contact, i.e., gas–liquid–solid three-phase boundary, the force (γ^l per unit length) acts on the boundary (vertical to the boundary and along the bubble surface). Since the force acting on the boundary of δl length is $\gamma^l\delta l$, the force along Cz is $-\gamma^l\cos\theta'\times\delta l$. In this case, the value becomes negative because the surface tension acts in the reverse direction with respect to the z-axis. Since the length of the boundary line is $2\pi d$ and the surface tension γ^l acts equally on the boundary line, the total force acting in the z-axis direction, which is caused by the surface tension, amounts to $-\gamma^l\cos\theta'2\pi d$. Since $\cos\theta'=d/r$, the equation $|-\gamma^l\cos\theta'\cdot 2\pi d|=2\gamma^l/r\cdot\pi d^2$ is derived. Since the force exerted by the pressure difference $p^\alpha-p^\beta$ and that exerted by the surface tension are balanced and remain stationary: $\left(p^\alpha - p^\beta\right)\pi d^2 - (2\gamma^l/r)\pi d^2 = 0$. Therefore:

$$p^\alpha - p^\beta = \frac{2\gamma}{r}. \qquad (2.28)$$

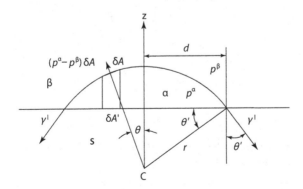

FIGURE 2.7 Mechanical balance for a case in which a part of the bubble (lens shape) is attached to the solid plane under zero gravity.

This equation is Laplace's equation (also known as the Young–Laplace equation or Laplace's formula) for spherical interfaces.*

If the interface is not spherical, it can be expressed using the following generalized form:

$$p^{\alpha} - p^{\beta} = \gamma \left(\frac{1}{r_1} + \frac{1}{r_2} \right), \tag{2.61}$$

where r_1 and r_2 are the main radii of curvature of the curved surface (see Figure 2.8).

Laplace's equation can also be derived from the thermodynamic theory. The derivation process of Equation (2.61) is omitted here, but please refer to reference 2 for details.

2.3.3 MARANGONI EFFECT

Thus far, a case in which the system at the thermodynamic equilibrium state is maintained at mechanically static equilibrium state has been described. However, if the system is in a thermodynamic non-equilibrium state, how would the mechanical treatment of the system be?

Surface tension and interfacial tension change depending on the temperature and composition of the system, or in some cases, the voltage applied to the interface. Therefore, if the surface–interfacial tension at a liquid surface or liquid–liquid interface locally changes due to the changes in the above factors, the motion of the liquid phase changes because of the action of the tangential force corresponding to the surface–interfacial tension difference (see Figure 2.9). The local change in the surface–interfacial tension in such a case is called the Marangoni effect. More generally, the Marangoni effect is the "local change in surface–interfacial tension involved in a dynamical plane in fluid mechanics"[9] corresponding to each specific case.

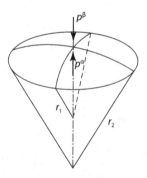

FIGURE 2.8 Pressure inside the interface p^{α} and pressure outside the interface p^{β} with principal radii of curvature radii r_1 and r_2.

---

* Because equations (2.28) and (2.61) generally hold not only for the gas–liquid interface but also for other interfaces such as the liquid/liquid interface, the surface/interfacial tension is simply denoted as γ without specifying the phase.

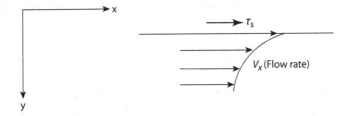

FIGURE 2.9 Liquid motion induced by surface shear stress τ_s because of the surface tension gradient $d\gamma/dx$ in the x-axis direction.

The local change in surface–interfacial tension γ is caused by local variations such as temperature T and concentration c of surfactant solute, or in some cases, by electrochemical factors.

One such significant occurrence is known as the "tears of wine" phenomenon. When a relatively strong wine in terms of alcohol content is poured into a glass, the wine goes up along the inner wall, and rows of droplets emerge. It was first explained by J. Thomson from the UK in 1855 as a phenomenon generated by the difference in the local surface tension as a driving force.[10]

In other words, the "tears of wine" are induced by the Marangoni effect originating from a concentration gradient. The alcohol in the liquid film on the inner wall evaporates, and the surface tension increases. As a result, the wine below is pulled against gravity by the Marangoni effect. The explanation provided by Thomson was basically correct, but for some reason it was forgotten. Now, this phenomenon is named after the Italian physicist Marangoni, who claimed the priority with the paper published in 1871,[11] 16 years after Thompson's first explanation.

In a system where the temperature gradient, concentration gradient, or electrical capillarity exists along the x-direction of the interface, if there is an electric potential gradient ψ, the surface–interfacial shear stress τ_s induced by the surface–interfacial tension gradient is expressed as:

$$\tau_s = \frac{d\gamma}{dx} = \frac{\partial\gamma}{\partial T}\cdot\frac{dT}{dx} + \frac{\partial\gamma}{\partial c}\cdot\frac{dc}{dx} + \frac{\partial\gamma}{\partial\psi}\cdot\frac{d\psi}{dx}. \qquad (2.62)$$

The liquid flow generated by this shear stress is obtained by solving the equations of motion (Navier–Stokes equations) and thermal conduction equation or diffusion equation under the boundary condition that gives the surface–interfacial tangential force.

As a dimensionless number characterizing the flow generated by the Marangoni effect, i.e., the Marangoni flow, the Marangoni number Ma is used. The Marangoni convection generated by the difference in the surface–interfacial tension (i.e., the surface–interfacial tension gradient) originating from the temperature difference (or temperature gradient) is characterized by Equation (2.63) as the ratio of the heat quantity transferred accompanied with the flow that is a mechanical motion of a liquid and the heat quantity transferred by heat conduction, which is a heat transport phenomenon:

$$Ma = \frac{\partial \gamma}{\partial T} \cdot \Delta T \cdot \frac{L}{a\eta},$$ (2.63)

where ΔT is the temperature difference, L is the characteristic length, a is the thermal diffusivity, and η is the viscosity. The value of Ma also represents the tendency according to which Marangoni convection is spontaneously generated in the system. For the Marangoni convection generated by the difference in the surface–interfacial tension due to the concentration difference, the value of Ma is also expressed by the following equation:

$$Ma = \frac{\partial \gamma}{\partial c} \cdot \Delta c \cdot \frac{L}{D\eta},$$ (2.64)

where Δc is the concentration difference, and D is the diffusion coefficient.

Among the different Marangoni flows, the surface–interfacial disturbance is the most characteristic. A well-known example is a case in which a surface-active component, e.g., ether, is evaporated from the surface of an aqueous solution. As shown in Figure 2.10,[12] when turbulence B, such as a vortex, is generated in the air flow near the surface, the difference in the concentration of ether evolves at the surface of the aqueous solution because the supply rate of ether through diffusion from the bulk phase to B' is extremely small compared to the supply rate to B' through surface flow and the vaporization rate from B'. Then, the region B' where the concentration becomes low, pulls the high-concentration region A', which begins to spread. As a result, the surface tension of A' is further lowered because the solution with higher ether concentration in the bulk solution is supplied to the surface A' region. The spreading motion is enhanced, and vortex A is generated. This is the generation of surface disturbance.*

Sternling et al.[14] analyzed such instability of the system caused by the Marangoni effect using the linear theory. However, the experimental results cannot be fully

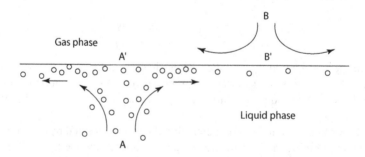

FIGURE 2.10 Mechanism of interfacial disturbance caused by the Marangoni effect. (Adapted from reference 12.)

---

* On the other hand, when the surfactant is enriched as the distance from the source point increases by, for example, the adsorption of surfactant at the interface, the flow at the interface is hindered.[13]

explained by their approach. In principle, such an instability problem should be treated as a nonlinear instability system. This field currently belongs to the most advanced field of physics, and future progress in elucidating the instability is awaited.

2.4 INTERFACIAL PHENOMENA AT EQUILIBRIUM

In this section, the static interfacial phenomena are taken up, and the understanding of them based on the "fundamentals of treating the interface" discussed in the previous sections is deepened. The static interfacial phenomenon refers to a thermodynamic equilibrium state or an interfacial phenomenon when a static balance is mechanically maintained.

2.4.1 ADSORPTION

As already mentioned, the excess Helmholtz energy f^s exists at the surface, and in liquid solution systems having two or more components certain phenomena occur in which the concentration distribution of the components differs between the bulk and surface phases. Such a concentration difference between bulk and surface phases is generally called adsorption. The case in which a certain component is in excess at the surface is referred to as positive adsorption, while the one in which a certain component is deficient is referred to as negative adsorption. The adsorption phenomenon is considered to occur irrespective of the equilibrium or non-equilibrium states of the system. However, as to the quantitative treatment, the method based on Gibbs' adsorption equation at the equilibrium state described next is still used as the only useful method.

2.4.1.1 Gibbs' Adsorption Equation

The basic equation on internal energy for a homogeneous open system is:

$$dU = TdS - pdV + \sum_{i=1}^{r} \mu_i dn_i. \tag{2.65}$$

From the discussion so far, the term $-PdV$ is substituted with γdA at a surface. At equilibrium, since $\mu_i^\alpha = \mu_i^\beta = \mu_i^s$ (Equation (2.17)):

$$dU^s = TdS^s + \gamma dA + \sum_{i=1}^{r} \mu_i dn_i^s. \tag{2.66}$$

Also, in comparison with:

$$dF^s = \gamma dA - S^s dT, \tag{2.12}$$

G^s is can be defined as:

$$G^s = U^s - TS^s - \gamma A. \tag{2.67}$$

Therefore, dG^s is:

$$dG^s = dU^s - TdS^s - S^sdT - \gamma dA - Ad\gamma,$$

$$= -S^sdT + \sum_{i=1}^{r} \mu_i dn_i^s - Ad\gamma \tag{2.68}$$

G^s is an extensive variable, and it was found experimentally that G^s is given by a linear homogeneous function of n_i^s under the condition where T and γ are constant. This means that:

$$G^s = \sum_{i=1}^{r} \mu_i n_i^s. \tag{2.69}$$

Therefore:

$$dG^s = \sum_{i=1}^{r} \left(\mu_i dn_i^s + n_i^s d\mu_i \right). \tag{2.70}$$

Since equations (2.68) and (2.70) are equivalent:

$$S^sdT + \sum_{i=1}^{r} n_i^s d\mu_i + Ad\gamma = 0. \tag{2.71}$$

Dividing both sides of Equation (2.71) by A, the following is obtained:

$$s^sdT + \sum_{i=1}^{r} \Gamma_i d\mu_i + d\gamma = 0. \tag{2.72}$$

This equation corresponds to the Gibbs–Duhem equation for the bulk phase, which is called the generalized Gibbs' adsorption equation.

Under the condition where T is constant:

$$d\gamma = -\sum_{i=1}^{r} \Gamma_i d\mu_i. \tag{2.73}$$

When the above equation is applied to the binary system:

$$d\gamma = -\Gamma_1 d\mu_1 - \Gamma_2 d\mu_2. \tag{2.74}$$

At the dividing surface S where $\Gamma_1 = 0$:

$$d\gamma = -\Gamma_{2(1)} d\mu_2. \tag{2.75}$$

This equation was first derived by Gibbs and is the so-called Gibbs' adsorption equation. Since $d\mu_2 = RTd \ln a_2$, Equation (2.75) is expressed with the following

Equation (2.76), or $\Gamma_{2(1)}$ indicates the value of Γ_2 at the dividing surface S where $\Gamma_1 = 0$, as shown in Figure 2.11.

$$\Gamma_{2(1)} = -\left(\frac{\partial \gamma}{\partial \mu_2}\right)_T = -\frac{a_2}{RT}\left(\frac{\partial \gamma}{\partial a_2}\right)_T, \tag{2.76}$$

where a_2 is the activity of component 2. If the concentration of component 2, c_2, is so low that it can be regarded as an ideal dilute solution, Equation (2.76) is rewritten as:

$$\Gamma_{2(1)} = -\frac{c_2}{RT}\left(\frac{\partial \gamma}{\partial c_2}\right)_T. \tag{2.77}$$

From the above equation, $\Gamma_{2(1)}$ can be obtained by measuring the relation between the surface tension and the activity or concentration of component 2. When the activity (or concentration) of the interfacially active components are the same, the larger is the ratio of decrease in surface tension to the increase in the activity (or concentration) of the interfacially active component, the larger is $\Gamma_{2(1)}$.*

Then, how can the difference in the concentration distribution of surface-active components between the surface and bulk, when a system containing the adsorbed surface is in a thermodynamic equilibrium state, be understood? Since the system is in an equilibrium state, the relationship between the chemical potential of the α and β phases and the surface is given as follows (Equation (2.17)):

$$\mu_i^\alpha = \mu_i^\beta = \mu_i^s. \tag{2.17}$$

μ_i^α and μ_i^β can be expressed as follows, respectively:

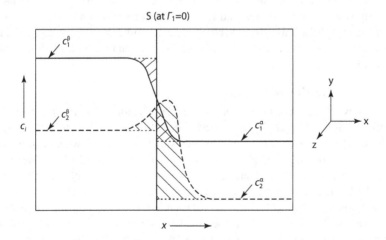

FIGURE 2.11   Adsorption amount of component 2 at the dividing surface where $\Gamma_1 = 0$.

---

* It was experimentally confirmed that the Gibbs' adsorption equation holds.[15]

$$\mu_i^\alpha = \mu_i^{\alpha,\circ} + RT \ln a_i^\alpha, \tag{2.78}$$

$$\mu_i^\beta = \mu_i^{\beta,\circ} + RT \ln a_i^\beta, \tag{2.79}$$

where $\mu_i^{\alpha,\circ}$ and $\mu_i^{\beta,\circ}$ are the chemical potential at the standard state for the respective phases. If the standard state of μ_i^s is chosen the same as that of μ_i^α, i.e., $\mu_i^{\alpha,\circ}$:

$$\mu_i^s = \mu_i^{\alpha,\circ} + RT \ln a_i^s = \mu_i^{\alpha,\circ} + RT \ln\left(\gamma_i^s \cdot x_i^s\right), \tag{2.80}$$

where γ_i^s is the activity coefficient when the standard state is chosen as $\mu_i^{\alpha,\circ}$ and the concentration unit is chosen as x_i (mole fraction). Although $x_i^s > x_i^\alpha$ for the positive adsorption, $a_i^s = a_i^\alpha$ because the standard states of the activities are the same. Therefore, it should be $\gamma_i^s = \gamma_i^\alpha$. In other words, in systems where adsorption occurs, it can be interpreted that the activity coefficients of the components are different between the surface and bulk.

2.4.2 WETTING

The term "wetting" as an interfacial phenomenon is generally used to mean "the liquid phase makes contacts with a solid surface and covers it." The phenomenon of wetting is an important subject widely related to general engineering, mainly in materials engineering. In the discussion on wetting, there has not been sufficient systematic description yet because 1) the solid phase is involved in the phenomenon and 2) the chemical reaction occurs to a greater or lesser extent at the contact between the solid and liquid phases.

Therefore, in this section, the system where (1) the solid phase surface is atomically smooth and uniform and (2) only the formation and extinction of the interface occur at the solid–liquid interface, while other reactions, e.g., chemical reaction, do not occur, and the system is in thermodynamic equilibrium state, is considered.

2.4.2.1 Classification of Wetting

When "get wet" is referred to as an interfacial phenomenon, the most common example is that (1) the liquid phase spreads in a thin-film-like state on a solid surface (Figure 2.12a). However, it can also "get wet" in a situation where (2) the liquid phase permeates into a paper, cloth, or porous material (Figure 2.12b). When (3) the liquid phase becomes a droplet state and is partially in contact with the solid surface (Figure 2.12c), the contact part can also be regarded as "wet." The wetting in the above-mentioned situations (1), (2), and (3) is called spreading wetting, immersional wetting, and adhesional wetting, respectively. All cases of wetting are common in that the phenomenon involves changes where the gas–solid interface disappears, and the solid–liquid interface is generated. Since gas is generally adsorbed on a solid surface, it can be said that the above process of change is actually the phenomenon in which the liquid phase expels gas from the solid surface.

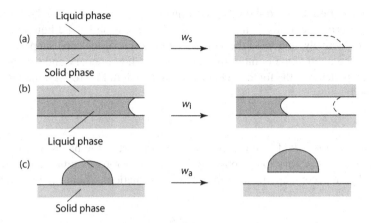

FIGURE 2.12   Change in interfacial free energy accompanied by various types of wettings.

2.4.2.2   Measure of Wetting

As a measure that instinctively expresses the degree of wetting, the angle θ between a droplet and a solid surface, i.e., contact angle, shown in Figure 2.13, is generally used. In fact, the degree of θ often plays a major role in many wetting phenomena. Meanwhile, as a general measure that is useful for comparing the wetting in various systems, the amount of change in the interfacial free energy on wetting is often used. Referring to Figure 2.12, the amounts of changes in the interfacial free energy for the above-mentioned spreading, immersional, and adhesional wetting are given by equations (2.81), (2.82), and (2.83), respectively:

$$w_s = \gamma^s - \gamma^l - \gamma^{ls}, \tag{2.81}$$

$$w_i = \gamma^s - \gamma^{ls}, \tag{2.82}$$

$$w_a = \gamma^s + \gamma^l - \gamma^{ls}, \tag{2.83}$$

where w_s, w_i, and w_a are called the work of spreading, work of immersion, and work of adhesion, respectively. These quantities correspond to the work necessary to

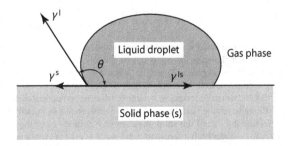

FIGURE 2.13   Contact angle at the solid–droplet interface, and surface (interfacial) tension.

retreat or separate the solid–liquid contact interface by a unit area for the respective wetting pattern. The terms γ^l and γ^s are the surface tension for the liquid and solid phases, respectively, and γ^{ls} is the liquid–solid interfacial tension. Suppose that γ^l, γ^s, and γ^{ls} are acting on a point at the gas–liquid–solid boundary and that the horizontal components of the force are balanced, as shown in Figure 2.13, the following is obtained:

$$\gamma^s = \gamma^{ls} + \gamma^l \cos\theta. \tag{2.84}$$

Equation (2.84) is called Young's equation.* In this case, the balance of the vertical force is not necessary to be considered because the force is supported by a solid surface (by an accompanying slightly elastic deformation). Then, the following are obtained:

$$w_s = \gamma^l (\cos\theta - 1), \tag{2.85}$$

$$w_i = \gamma^l \cos\theta, \tag{2.86}$$

$$w_a = \gamma^l (\cos\theta + 1). \tag{2.87}$$

Equation (2.87) is called the Young–Dupre equation.

From equations (2.85)–(2.87), w_s, w_i, and w_a can be obtained by measuring γ^l and θ. If $\theta \leqq 180°$, $w_a \geqq 0$, and adhesional wetting occurs spontaneously. If $\theta < 90°$, $w_i \geqq 0$, and immersional wetting occurs spontaneously. If $\theta = 0°$, $w_s = 0$, and the spreading wetting occurs spontaneously. Using the work of cohesion $w_c = 2\gamma^l$, Equation (2.87) is rewritten as:

$$\cos\theta = \frac{2w_a}{w_c} - 1, \tag{2.88}$$

---

* Several controversies are associated with the Young's equation with respect to its accuracy. It has not been experimentally confirmed because methods with a reasonable principle that can be used to measure the interfacial tension including the solid phase, such as γ^s and γ^{ls} in Young's equation, have not yet been developed. Recently, Boruvka and Neumann (1977) derived the following equation for a strict equilibrium contact angle:

$$\gamma^{ls} - \gamma^s + \gamma^l \cos\theta + \tau\kappa = 0, \tag{2.89}$$

where τ denotes the line tension and κ denotes the curvature of the border. As for the value of τ, Petheca (1977) estimated that the value of τ/γ^l is approximately 10^{-9} m at the interface between the ordinary liquid phase and the solid phase. Therefore, when a droplet having a curvature greater than the order of μm is considered, the aforementioned equation can be considered to be almost equivalent to Young's Equation (2.84). For the descriptions on line tension and other related concepts, reference 3 was consulted.It is difficult to ensure that the solid surface achieves an atomically flat state described at the beginning of Section 2.4.2, and the solid phase surface exhibits some degree of roughness. In addition, it is necessary to consider both microscopic and macroscopic unevenness and the shape of the interface specific to the solid phase. Therefore, Young's equation (2.84) should be applied to the real system with extra care. This is specifically described in Sections 2.5.1.3, 3.1.2.4, and 3.3.3.

and θ is expressed as the function of w_a and w_c. Similarly, it is easily concluded that θ can be expressed as a function of w_s and w_c and w_i and w_c for equations (2.85) and (2.86).

2.4.2.3 Extension of the Wetting Concept

As described in the classification of wetting, wetting can be regarded as the state of existence of a liquid phase at a gas–solid interface. Let this concept be extended, and the state of existence of α and β be thermodynamically considered, which are two liquid phases, and foreign particle phase δ at the interface. Here, it is assumed that there are no chemical reactions among these phases. According to Figure 2.14,[16] $\Delta F_a^{\,s}$ and $\Delta F_b^{\,s}$ are defined by the following equations (2.90) and (2.91).

$$\Delta F_a^s = A^{i\alpha}\gamma^{\delta\alpha} + A^{i\beta}\gamma^{\delta\beta} - A^i\gamma^{\alpha\beta} - A^\alpha\gamma^{\delta\alpha}, \tag{2.90}$$

$$\Delta F_b^s = A^\beta\gamma^{\delta\beta} - \left(A^{i\alpha}\gamma^{\delta\alpha} + A^{i\beta}\gamma^{\delta\beta} - A^i\gamma^{\alpha\beta}\right), \tag{2.91}$$

where $\gamma^{\delta\alpha}$ and $\gamma^{\delta\beta}$ are the interfacial tension between the particle and the α or β phase, respectively; $\gamma^{\alpha\beta}$ is the interfacial tension between the α and β phases; A^α and A^β are the interfacial areas between the particle and the α or β phase; and A^i is the area of the region in which the α–β interface disappears through particle adhesion. The subscript "a" indicates the process in which the particles in the α phase are transferred to the α–β interface and adhere on it ((a) in Figure 2.14), and the subscript "b" indicates the process in which the adhered particles are transferred into the β phase ((b, i) or (b, ii) in Figure 2.14). $\Delta F_a^{\,s}$ and $\Delta F_b^{\,s}$ correspond to the changes in Helmholtz energy in processes (a) and (b), respectively.

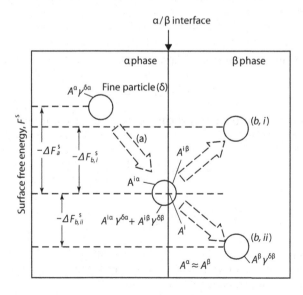

FIGURE 2.14 Thermodynamic illustration for the system of fine particles (δ) in α phase, at the α–β interface, or in β phase. (Adapted from reference 16.)

When the particles stably exist at the interface of the α–β phase, both $\Delta F_a^s < 0$ and $\Delta F_{b,i}^s > 0$ must hold simultaneously. When the particles migrate into the β phase through the interface and exist stably, both $\Delta F_a^s < 0$ and $\Delta F_{b,ii}^s < 0$ must hold simultaneously.

When the particles are identical to those of the β phase, i.e., the phase on the right-hand side, $\gamma^{\beta\beta} = 0$; thus, when the geometric relationship shown in Figure 2.14 is considered, it is easily understood that $\Delta F_a^s < 0$ and $\Delta F_b^s < 0$ holds. This means that the state in which the β-phase particles exist in the α phase (dispersion state, described later) is a thermodynamic non-equilibrium state. The particles in this state spontaneously coalesce and are eventually absorbed into the β phase.

2.4.3 EFFECT OF CURVATURE

When the interface of the α–β phases is not a flat plane, the pressure in the α phase becomes higher than that in the β phase, as shown in Laplace's equations (2.28) and (2.61). As a result, the chemical potential in the α phase increases, resulting in a change in the relation of the thermodynamic equilibrium state between the α and β phases. In what follows, the equilibrium vapor pressure and heat of vaporization for droplets and bubbles and the melting point and solubility of solid particles are specifically discussed.

2.4.3.1 Vapor Pressure

《Droplet》

A system composed of pure droplets with radius r (α phase) and the gas phase (β phase) is considered.

The mechanical equilibrium of the system is expressed by Laplace's Equation (2.28).

$$p^\alpha - p^\beta = \frac{2\gamma^{\alpha\beta}}{r}. \tag{2.28'}$$

The thermodynamic equilibrium is expressed as:

$$\mu^\alpha = \mu^\beta = \mu^s. \tag{2.17'}$$

Let it be considered that there is a small reversible change δ in the system. In this case:

$$\delta p^\alpha - \delta p^\beta = \delta\left(\frac{2\gamma^{\alpha\beta}}{r}\right), \tag{2.92}$$

$$\delta\mu^\alpha = \delta\mu^\beta = \delta\mu^s. \tag{2.93}$$

When the temperature is constant, the following equations (2.94) and (2.95), well known through molar volumes v^α and v^β, respectively, hold with respect to the relation between the chemical potential and pressure.

$$\delta\mu^\alpha = v^\alpha \delta p, \tag{2.94}$$

$$\delta\mu^{\beta} = v^{\beta}\delta p. \tag{2.95}$$

From equations (2.93)–(2.95):

$$v^{\alpha}\delta p^{\alpha} = v^{\beta}\delta p^{\beta}. \tag{2.33}$$

Substituting Equation (2.33) into Equation (2.92):

$$\delta\left(\frac{2\gamma^{\alpha\beta}}{r}\right) = \frac{v^{\beta} - v^{\alpha}}{v^{\alpha}}\delta p^{\beta}, \tag{2.96}$$

$$\delta\left(\frac{2\gamma^{\alpha\beta}}{r}\right) = \frac{v^{\beta} - v^{\alpha}}{v^{\beta}}\delta p^{\alpha}. \tag{2.97}$$

As already described, the α phase is a droplet, and the β phase is a gas phase. In this case, $v^{g} \gg v^{l}$, v^{l} can be ignored. Furthermore, considering that the equilibrium vapor pressure of the liquid phase is around 1 atm at maximum, the gas can be treated as an ideal gas; therefore:

$$v^{g} = \frac{RT}{p^{g}}, \tag{2.98}$$

where R is the gas constant. Hereafter, the term p^{g} is written as p for simplicity. Substituting Equation (2.98) into Equation (2.96), the following is obtained:

$$\delta\left(\frac{2\gamma^{l}}{r}\right) = \frac{RT}{v^{l}}\frac{\delta p}{p}. \tag{2.99}$$

The relation $v^{g} \gg v^{l}$ is also applied here. When the value of r approaches infinity, $1/r = 0$, and $p = p_{\circ}$. Therefore, by integrating Equation (2.99) from $p_{\circ}$ to p_{r} (vapor pressure at $1/r$), the following Equation (2.100) is obtained:[*]

$$\ln\frac{p_{r}}{p_{\circ}} = \frac{2\gamma^{l}}{r}\frac{v^{l}}{RT}. \tag{2.100}$$

---

[*] When p is high and v^{l} cannot be ignored in comparison with v^{g}, Equation (2.100) should be rewritten as follows:

$$\delta\left(\frac{2\gamma^{l}}{r}\right) = \frac{v^{g}}{v^{l}}\delta p - \delta p. \tag{2.101}$$

When the above equation is integrated, we obtain:

$$\frac{2\gamma^{l}}{r} = \frac{RT}{v}\ln\frac{p_{r}}{p_{\circ}} - (p_{r} - p_{\circ}). \tag{2.102}$$

The above equation is a more accurate expression.

Herein, v^l is assumed to be constant. The above Equation (2.100) is well known as Kelvin's equation (see Figure 2.15).

According to Kelvin's equation, the smaller the droplet, the larger is the vapor pressure equilibrated with the droplet.

《**Bubble**》

For bubbles, the α phase is the gas phase, and the β phase is the liquid phase. By assuming that v^l can be ignored since $v^g \gg v^l$, Equation (2.97) becomes:

$$\delta\left(\frac{2\gamma^l}{r}\right) = -\frac{RT}{v^l}\frac{\delta p}{p}. \tag{2.103}$$

Herein, the relation $v^g = RT/p$ is applied. By integrating the above equation:[*]

$$\ln\frac{p_r}{p_\circ} = -\frac{2\gamma^l}{r}\frac{v^l}{RT}. \tag{2.104}$$

From the above equation, it is understood that the smaller the bubble in the liquid, the smaller the vapor pressure in the bubble.

2.4.3.2 Heat of Vaporization

Let's consider the surface of a pure droplet, specifically the curved surface at the upper end of the liquid phase in a capillary (described below), with radius r at temperature T. If the wettability between the capillary and droplet is bad ($\theta > 90°$), the curved surface at the upper end can be regarded as a part of a droplet with radius r. If it is supposed that the evaporated liquid is always supplied through the capillary and

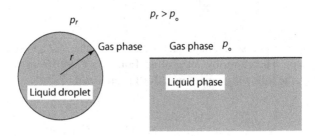

FIGURE 2.15 Equilibrium vapor pressure p_r of a droplet having radius r and the equilibrium vapor pressure p of the flat liquid phase ($r \rightarrow \infty$).

---

[*] Similar to Equation (2.102), a more accurate expression of Equation (2.104) can be obtained by integrating Equation (2.97) without ignoring the value of v^β, i.e.,

$$\delta\left(\frac{2\gamma^l}{r}\right) = \delta p^\alpha - \frac{v^\alpha}{v^\beta}\delta p^\alpha. \tag{2.105}$$

When the above equation is integrated, we obtain:

$$\frac{2\gamma^l}{r} = (p_r - p_\circ) - \frac{RT}{v^l}\ln\frac{p_r}{p_\circ}. \tag{2.106}$$

that the temperature of the system is kept constant, the process can be regarded as reversible vaporization at constant T and r. To keep the system at a constant temperature, the heat dQ (see the following Equation (2.107) accompanied by vaporization needs to be supplied.

$$dQ = TdS. \tag{2.107}$$

This vaporization process is a reversible process under the constant T, r, surface area of the droplet, and surface state. Therefore, the entropy of the surface phase is constant, and s^g and s^l are also constant. Consequently, the entropy change dS of the system accompanied by the vaporization of dn mole of liquid is given by the following Equation (2.108).

$$dS = \left(s^g - s^l\right)dn. \tag{2.108}$$

Therefore, the heat of vaporization Δh_{vap} accompanied by the vaporization of one mole (molar heat of vaporization) is expressed as:

$$\Delta h_{vap} = T\left(s^g - s^l\right). \tag{2.109}$$

Since the degree of freedom of this system is 2 (see Equation (2.138)), Δh_{vap} depends on T and r. Then, let's consider the dependence of the radius on Δh_{vap}. s^g depends on T and p^g, and s^l depends on T and p^l. Under the condition where T is constant:

$$\delta\Delta h_{vap} = T\left(\frac{\partial s^g}{\partial p^g}\delta p^g - \frac{\partial s^l}{\partial p^l}\delta p^l\right). \tag{2.110}$$

Using cross-differentiation:

$$\left(\frac{\partial s}{\partial p}\right)_r = \left[\frac{\partial\left\{-\left(\frac{\partial\mu}{\partial T}\right)_p\right\}}{\partial p}\right]_T = \left[\frac{\partial\left\{-\left(\frac{\partial\mu}{\partial p}\right)_T\right\}}{\partial T}\right]_p = -\left(\frac{\partial v}{\partial T}\right)_p. \tag{2.111}$$

Therefore, Equation (2.110) can be written using equations (2.96) and (2.97) as follows:

$$\delta\Delta h_{vap} = -T\left(\frac{\partial v^g}{\partial T}\cdot\frac{v^l}{v^g - v^l} - \frac{\partial v^l}{\partial T}\cdot\frac{v^g}{v^g - v^l}\right)\delta\left(\frac{2\gamma^l}{r}\right). \tag{2.112}$$

Assuming ideal gas condition:

$$v^g = \frac{RT}{p^g} \quad \left(\frac{\partial v^g}{\partial T}\right)_{p^g} = \frac{R}{p^g}. \tag{2.113}$$

In addition, assuming that v^l is negligible since $v^l \ll v^g$, Equation (2.112) is expressed as:

$$\delta \Delta h_{vap} = -\left\{ v^l - T\left(\frac{\partial v^l}{\partial T}\right)_{p^l} \right\} \delta\left(\frac{2\gamma^l}{r}\right). \tag{2.114}$$

By integrating Equation (2.114) from the flat surface ($r \to \infty$) to the droplet with radius r under the constant T:

$$\Delta h_{vap,r} - \Delta h_{vap,\circ} = -\frac{2\gamma^l}{r}\left\{ v^l - T\left(\frac{\partial v^l}{\partial T}\right)_{p^l} \right\}. \tag{2.115}$$

Since, in general, it can be considered as $v^l \gg T(\partial v^l / \partial T)_{p^l}$, Equation (2.115) is expressed as:

$$\Delta h_{vap,r} - \Delta h_{vap,\circ} = -\frac{2\gamma^l v^l}{r}. \tag{2.116}$$

From equations (2.115) and (2.116), it is understood that the heat of vaporization $\Delta h_{vap,r}$ for a droplet with radius r is smaller than $\Delta h_{vap,\circ}$ for a flat surface. However, the degree of this decrease is so small that the decrease in the heat of vaporization is only 0.6% for water at 4°C even if r is 10 nm.

2.4.3.3 Melting Point

Let the relation between the curvature of a solid–liquid interface (pure component) and its melting point, known as the Gibbs–Thomson effect be considered.

When both sides of the Gibbs–Duhem Equation (2.32) for the α and β phases are subtracted from each other under constant pressure in the β phase, since $\delta\mu^\alpha = \delta\mu^\beta$:

$$\Delta s \delta T + v^\alpha \delta p^\alpha = 0, \tag{2.117}$$

where $\Delta s = s^\beta - s^\alpha$. When the α phase is solid, and the β phase is liquid, Δs corresponds to $\Delta_f s$, i.e., the entropy of fusion. In addition, from Equation (2.30), which is the differential form of Laplace's equation, and $\delta p^l = 0$,

$$\delta p^s = \delta\left(\frac{2\gamma^{ls}}{r}\right). \tag{2.118}$$

At equilibrium:

$$\Delta_f \mu = \mu^l - \mu^s = \Delta_f h - \Delta_f s T_m = 0, \tag{2.119}$$

where T_m is the melting point. Therefore:

$$\Delta_f s = \frac{\Delta_f h}{T_m}. \tag{2.120}$$

Consequently, Equation (2.117) becomes:

$$\frac{\delta T}{T_m} = -\frac{v^s}{\Delta_f h} \delta\left(\frac{2\gamma^{ls}}{r}\right).$$

(2.121)

When it is assumed that the value of $\Delta_f h$ (enthalpy of fusion) is constant regardless of the curvature and that v^s is also constant, and the above Equation (2.121) is integrated from $1/r=0$ (flat surface) to $1/r$, the following Equation (2.122) is obtained:

$$\ln\frac{T_{m,r}}{T_{m,\circ}} = -\frac{2\gamma^{ls}}{r} \cdot \frac{v^s}{\Delta_f h}.$$

(2.122)

The above equation is called Thomson's equation, where $T_{m,r}$ is the melting point at the solid–liquid interface with radius r and $T_{m,\circ}$ is the melting point at the solid–liquid interface at $r \to \infty$, i.e., the normal melting point for a flat surface. Since all variables quantities on the right side of Equation (2.122) are positive, $\ln(T_{m,r}/T_{m,\circ}) < 0$, and thus, $T_{m,r} < T_{m,\circ}$.

Whether the solid–liquid interface has a smooth curved surface is a matter of contention, but at any rate Equation (2.122) is generally accepted as an explanation for a change in the melting point of a solid–liquid interface having different curved surfaces.

2.4.3.4 Solubility

The Gibbs–Duhem equation for the r-component system is:

$$S dT - V dp + \sum_{i=1}^{r} n_i d\mu_i = 0.$$

(2.123)

Dividing both sides of the equation by the volume of the system:

$$\bar{s} dT - dp + \sum_{i=1}^{r} c_i d\mu_i = 0,$$

(2.124)

where $\bar{s}$ is an entropy per unit volume. When the temperature is constant in Equation (2.124):

$$dp = \sum_{i=1}^{r} c_i d\mu_i.$$

(2.125)

Substituting the differential form of Laplace's Equation (2.30) into Equation (2.125) for the α and β phases, the following is obtained:

$$d\left(\frac{2\gamma^{\alpha\beta}}{r}\right) = \sum_{i=1}^{r}\left(c_i^\alpha - c_i^\beta\right) d\mu_i.$$

(2.126)

Now, the α phase is defined as a solid phase, the β phase as a liquid phase, and the solid phase is assumed to be spherical. By assuming that the pressure p^l in the liquid phase is constant, i.e., $dp^l=0$, the following equation is obtained from Equation (2.125):

$$dp^l = \sum_{i=1}^{r} c_i^l d\mu_i = 0. \tag{2.127}$$

Therefore, Equation (2.126) is expressed as:

$$d\left(\frac{2\gamma^{ls}}{r}\right) = \sum_{i=1}^{r} c_i^s d\mu_i. \tag{2.128}$$

Now, the solid phase is assumed to be composed of a pure constituent 1, namely the values of $c_2^s, \ldots c_q^s$ are zero. In this case, Equation (2.128) becomes:

$$d\left(\frac{2\gamma^{ls}}{r}\right) = c_1^s d\mu_1. \tag{2.129}$$

Meanwhile:

$$\mu_1 = \mu_1^\circ + RT \ln \gamma_1 x_1. \tag{2.130}$$

Therefore, Equation (2.129) can be written as:

$$d\left(\frac{2\gamma^{ls}}{r}\right) = c_1^s RT d \ln\left(\gamma_1 x_1\right). \tag{2.131}$$

Since $c_1^s = 1/v_1^s = 1/v_1^0$, and v_1^0 is the molar volume of the pure solid phase 1:

$$d\left(\frac{2\gamma^{ls}}{r}\right) = \frac{RT}{v_1^\circ} d \ln\left(\gamma_1 x_1\right). \tag{2.132}$$

By integrating Equation (2.132) from $1/r=0$ (flat plane) to $1/r$:

$$\frac{2\gamma^{ls}}{r} = \frac{RT}{v_1^\circ} \ln\left(\frac{\gamma_{1,r} x_{1,r}}{\gamma_{1,\circ} x_{1,\circ}}\right). \tag{2.133}$$

In a case where the liquid phase is an ideal solution, Equation (2.133) can be written as:

$$\frac{2\gamma^{ls}}{r} = \frac{RT}{v_1^\circ} \ln\left(\frac{x_{1,r}}{x_{1,\circ}}\right). \tag{2.134}$$

This equation is the well-known Freundlich–Ostwald equation (see Figure 2.16).

If it is assumed that the surface tension does not depend on (or exist within the scope where it does not depend on) the size of the solid particle, the solubility of a solid composed of the pure component 1 increases with the decrease in the particle size based on Equation (2.133) (or Equation (2.134). Therefore, when solid particles with different sizes coexist in a solution simultaneously, large particles grow by

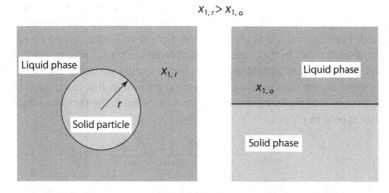

$$x_{1,r} > x_{1,o}$$

FIGURE 2.16 Solubility $x_{1,r}$ for a solid particle (pure component 1) having radius r and solubility $x_{1,o}$ for a flat solid–liquid interface ($r \to \infty$).

incorporating small particles, and angular particles become rounded. This phenomenon is known as Ostwald ripening (see Figure 2.17).

2.4.3.5 Phase Rule

As described in Sections 2.4.3.1–2.4.3.1.4, the properties of the system containing an interface depends on the curvature of such interface. From this observation, the degrees of freedom f of the system containing the interface are considered different from those derived from the following Equation (2.135) that considers only the bulk phase.

$$f = c + 2 - v - (q + \chi), \tag{2.135}$$

where c is the number of independent components, v is the number of phases, q is the number of the independent reactions, and χ is the number of concentration restriction accompanied by the reaction.

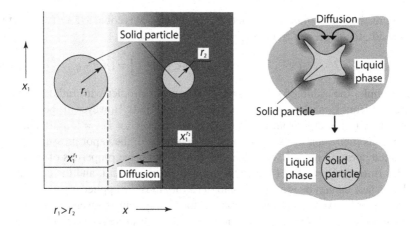

FIGURE 2.17 Ostwald growth (ripening).

Let a system composed of c number of independent components, v number of bulk phases, ξ types of interfaces, and q number of independent reactions now be considered. For simplicity, it is assumed that there is only one type of phase in the interface of each type. In this case, the number of interfacial phases is equal to ξ. If this system is in a non-equilibrium state, the following variables are necessary to perfectly describe the system using intensive properties. In this case, since the concentration is given as the number of moles per unit volume, the pressure p or volume V does not appear in the list.

$$\underbrace{T}_{\text{1 variable}} ; \underbrace{c_1^1, \cdots c_c^v}_{cv \text{ variables}} ; \underbrace{\Gamma_1^1, \cdots \Gamma_c^\xi}_{c\xi \text{ variables}} ; \underbrace{\gamma^1, \cdots \gamma^\xi}_{\xi \text{ variables}}. \tag{2.136}$$

Namely, the number of variables is equal to $1 + cv + c\xi + \xi$. If the system is in a thermodynamic equilibrium state, the following restrictions are added.

(1) Mechanical equilibrium: There are ξ number of restrictions because the Laplace's equation in Equation (2.28) holds for the ξ types of interfaces.
(2) The chemical potential of each component between each phase and interface is equal. Therefore, with respect to the relation between v number of phases and ξ types of the interfaces between the phases, the following relation holds for each component.

$$\mu_c^1 = \mu_c^{s(1/2)} = \mu_c^2 \cdots = \mu_c^{s(\xi)} = \mu_c^v, \tag{2.137}$$

where $s(1/2)$ indicates the interface between phases 1 and 2. In the above equation, the number of independent equations is $v + \xi - 1$. Thus, the number of restrictions is equal to $c(v + \xi - 1)$.
(3) The number of systems where the relation $\Delta G_q = 0$ holds is equal to the number of independent reactions. ΔG_q is the change in free energy of reaction q. Herein, it is assumed that there are not χ number of restrictions of the concentration associated with a chemical reaction.

Therefore, the total number of restrictions (1), (2), and (3) is equal to $\xi + c(v + \xi - 1) + q$. As a result, the degree of freedom f is:

$$f = 1 + cv + c\xi + \xi - \{\xi + c(v + \xi - 1) + q\} = 1 + c - q. \tag{2.138}$$

For example, the value of f for a system of a water droplet with radius r and gas phase is 2 from Equation (2.138), because $c = 1$ and $q = 0$. However, if the value of r is infinity, f becomes 1 from Equation (2.135), because $c = 1$, $v = 2$, $q = 0$, and $\chi = 0$. This means that when the temperature T is determined, the vapor pressure of water at this temperature has a fixed value and that all intensive properties of the system are determined. However, if the water droplet becomes small, and the contribution of the surface cannot be ignored, Equation (2.138) must be applied, resulting in $f = 2$. In other words, even if the temperature T for this system is given, its vapor pressure cannot be determined unless one more intensive property such as radius r is specified.

2.4.4 NUCLEATION

2.4.4.1 Homogeneous Nucleation

Let the changes in Helmholtz energy F be considered when the droplet α is formed in the β phase (gas or liquid phase). Since the free energy per unit interfacial area is expressed by Equation (2.15), the free energy F^s for the system with interfacial area A is:

$$F^s = \gamma^{\alpha\beta} A + \sum_{i=1}^{r} n_i^s \mu_i^s. \tag{2.139}$$

The free energies of the α and β phases are given by equations (2.140) and (2.141), respectively.

$$F^{\alpha} = \sum_{i=1}^{r} n_i^{\alpha} \mu_i^{\alpha} - p^{\alpha} V^{\alpha}. \tag{2.140}$$

$$F^{\beta} = \sum_{i=1}^{r} n_i^{\beta} \mu_i^{\beta} - p^{\beta} V^{\beta}. \tag{2.141}$$

Therefore, the total free energy F of the system containing a liquid phase is:

$$F = \sum_{i=1}^{r} n_i^{\alpha} \mu_i^{\alpha} + \sum_{i=1}^{r} n_i^{\beta} \mu_i^{\beta} + \sum_{i=1}^{r} n_i^s \mu_i^s - p^{\alpha} V^{\alpha} - p^{\beta} V^{\beta} + \gamma^{\alpha\beta} A. \tag{2.142}$$

Meanwhile, the free energy before the formation of the liquid phase, $F_{\circ}$, is given by

$$F_{\circ} = \sum_{i=1}^{r} n_i \mu_{i,\circ}^{\beta} - p_{\circ}^{\beta} V. \tag{2.143}$$

Herein, it is assumed that $n_i = n_i^{\alpha} + n_i^{\beta} + n_i^s$ and $V = V^{\alpha} + V^{\beta}$, i.e., the systems before and after the droplet formation are composed of the same number of atoms (molecules) and the same volume. The change in the free energy $(\Delta F)_{T,V}$ at the time of the formation of the droplet α at temperature T is given by subtracting Equation (2.143) from Equation (2.142) and is expressed by the following Equation (2.144).

$$\left(\Delta F \right)_{T,V} = F - F_{\circ}. \tag{2.144}$$

For simplicity, it is assumed that the β phase in the matrix phase is so large that the pressure and composition of the matrix phase do not change even if droplet α is formed.* In this case:

$$\mu_i^{\beta} = \mu_{i,\circ}^{\beta}, \quad p^{\beta} = p_{\circ}^{\beta}. \tag{2.145}$$

---

* Note that this assumption does not hold in some cases. In those cases, Equation (2.144) should be used. Please see the example described in Section 4.1.2.

Then, substituting equations (2.142), (2.143), and (2.145) into Equation (2.144), the following is obtained:

$$(\Delta F)_{T,V} = \sum_{i=1}^{r} n_i^\alpha \left(\mu_i^\alpha - \mu_i^\beta \right) + \sum_{i=1}^{r} n_i^s \left(\mu_i^s - \mu_i^\beta \right) - V^\alpha \left(p^\alpha - p^\beta \right) + \gamma^{\alpha\beta} A. \qquad (2.146)$$

Since the above Equation (2.146) is substantially derived under the constants T and V, the change in the free energy is hereafter expressed simply as ΔF by omitting the subscripts T and V. By using Laplace's Equation (2.28):

$$V^\alpha \left(p^\alpha - p^\beta \right) = \frac{4}{3} \pi r^3 \cdot \frac{2\gamma^{\alpha\beta}}{r} = \frac{2}{3} \left(4\pi r^2 \right) \gamma^{\alpha\beta} = \frac{2}{3} \gamma^{\alpha\beta} A. \qquad (2.147)$$

Therefore:

$$\Delta F = \sum_{i=1}^{r} n_i^\alpha \left(\mu_i^\alpha - \mu_i^\beta \right) + \sum_{i=1}^{r} n_i^s \left(\mu_i^s - \mu_i^\beta \right) + \frac{1}{3} \gamma^{\alpha\beta} A. \qquad (2.148)$$

If this system is in an equilibrium state, Equation (2.17), i.e., $\mu_i^\alpha = \mu_i^\beta = \mu_i^s$, holds. Therefore:

$$\Delta F_g = \frac{1}{3} \gamma_g^{\alpha\beta} A_g. \qquad (2.149)$$

The droplet in this case is the critical nucleus (denoted as g).

Next, let the changes in the free energy of the droplet (embryos) be considered to be smaller than the critical nucleus. In this case, since the system is not in an equilibrium state, the relation in Equation (2.17) cannot be applied. However, since it can be considered as $n_i^s \ll n_i^\alpha$ and $(\mu_i^s - \mu_i^\beta) < (\mu_i^\alpha - \mu_i^\beta)$, the term $n_i^s (\mu_i^s - \mu_i^\beta)$ can be ignored compared to the term $n_i^\alpha (\mu_i^\alpha - \mu_i^\beta)$. Then, Equation (2.148) becomes:

$$\Delta F = \sum_{i=1}^{r} n_i^\alpha \left[\mu_i^\alpha \left(T, p^\alpha, x_2^\alpha, \cdots, x_r^\alpha \right) - \mu_i^\beta \left(T, p^\beta, x_2^\beta, \cdots, x_r^\beta \right) \right] + \frac{1}{3} \gamma^{\alpha\beta} A. \qquad (2.150)$$

In Equation (2.150), the chemical potential is expressed as a function of T, p, and x_i. If changes in v_i^α by pressure can be neglected, Equation (2.152) is derived using Equation (2.151).

$$\left(\frac{\partial \mu_i^\alpha}{\partial p^\alpha} \right)_{T, x_i^\alpha} = v_i^\alpha \qquad (2.151)$$

$$\mu_i^\alpha \left(T, p^\alpha, x_2^\alpha, \cdots, x_r^\alpha \right) = \mu_i^\alpha \left(T, p^\beta, x_2^\alpha, \cdots, x_r^\alpha \right) + v_i^\alpha \left(p^\alpha - p^\beta \right). \qquad (2.152)$$

Substituting Equation (2.152) into Equation (2.150) and using the relation $V^\alpha = \sum_{i=1}^{r} n_i^\alpha v_i^\alpha$ and Equation (2.147), the following is obtained:

$$\Delta F = \sum_{i=1}^{r} n_i^\alpha \left[\mu_i^\alpha \left(T, p^\beta, x_2^\alpha, \cdots, x_r^\alpha \right) - \mu_i^\beta \left(T, p^\beta, x_2^\beta, \cdots, x_r^\beta \right) \right] + \gamma^{\alpha\beta} A. \quad (2.153)$$

The above equation indicates the value of ΔF, which is a more specific form of Equation (2.148).

Now, Equation (2.153) can be applied to a case in which a droplet of pure component 1 (α phase) nucleates. In this case, Equation (2.153) becomes:

$$\Delta F = n_1^\alpha \left[\mu_1^\alpha \left(T, p^\beta \right) - \mu_1^\beta \left(T, p^\beta, x_2^\beta, \cdots, x_r^\beta \right) \right] + \gamma^{\alpha\beta} A. \quad (2.154)$$

The term in brackets in Equation (2.154) represents the changes in the free energy per mole, $\Delta\mu_1$, when the phase transformation $\beta \rightarrow \alpha$ occurs with infinite radius. This value is relatively easy to obtain from a thermodynamics databook. Thus, Equation (2.154) becomes:

$$\Delta F = n_1^\alpha \Delta\mu_1 + \gamma^{\alpha\beta} A. \quad (2.155)$$

Since $n_1^\alpha = v^{\alpha'}/v_1^{\alpha,\circ}$, $v^{\alpha'} = 4/3\pi r^3$ (volume of the nucleus), and $A = 4\pi r^2$, Equation (2.155) becomes (see Figure 2.18):

$$\Delta F = \frac{4\pi r^3}{3v_1^{\alpha,\circ}} \Delta\mu_1 + 4\pi r^2 \gamma^{\alpha\beta}, \quad (2.156)$$

where $v_1^{\alpha,\circ}$ is the molar volume of the liquid phase of pure component 1 (α phase).

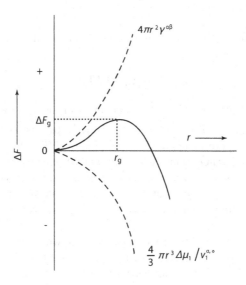

FIGURE 2.18 Relation between the free energy of formation of a sphere embryo ΔF and its radius r.

From the second law of thermodynamics, the following equation is derived at equilibrium, i.e., with critical nucleus:

$$(dF)_{T,V} = 0. \tag{2.157}$$

If the scenario where the droplet grows as a path of the change is considered:

$$\left(\frac{\partial F}{\partial r}\right)_{T,V} = \frac{4\pi r_g^2}{v_1^{\alpha,\circ}} \Delta\mu_1 + 8\pi r_g \gamma^{\alpha\beta} = 0. \tag{2.158}$$

Therefore:

$$r_g = -\frac{2v_1^{\alpha,\circ}\gamma^{\alpha\beta}}{\Delta\mu_1}. \tag{2.159}$$

When a pure droplet is nucleated from a gas phase:

$$\Delta\mu_1 = RT \ln\left(\frac{p_\circ^\circ}{p_r}\right), \tag{2.160}$$

where $p_\circ^\circ$ is the equilibrium vapor pressure for a pure droplet at $r=\infty$ and p_r is the vapor pressure of the gas phase in a supersaturated state. p_r is also the equilibrium vapor pressure for the droplet with radius r. Substituting Equation (2.160) into Equation (2.159), the following is obtained:

$$r_g = \frac{2v_1^{\alpha,\circ}\gamma^{\alpha\beta}}{RT \ln\left(\dfrac{p_r}{p_\circ^\circ}\right)}. \tag{2.161}$$

The above equation is Kelvin's equation. When a pure solid phase is nucleated from a liquid solution:

$$\Delta\mu_1 = -\Delta_m\mu_1^\circ - RT \ln \gamma_1 x_1, \tag{2.162}$$

where γ_1 is the activity coefficient of component 1 of the solution when a pure liquid is chosen as the standard state, and $\Delta_m\mu_1^\circ \left(= \mu_1^{\circ,l} - \mu_1^{\circ,s}\right)$ is the change in the free energy of fusion for pure component 1. Equation (2.160) corresponds to a case in which supersaturated water vapor becomes a water droplet, while Equation (2.162) corresponds to processes such as the nucleation of ice from seawater or that of nucleation of NaCl crystal from a supersaturated NaCl solution. With respect to the nucleation of solid, it is assumed that the surface tension of the crystal is uniform and that the crystal can be spherical.

Equation (2.146) can also be applied to the case in which a droplet of a chemical compound q or an n_q mole of solid particles is formed from a solution by chemical reaction. In this case, a very significant assumption is required that the volume change associated with the reaction must be ignored. Using the changes in the free energy per mole of the reaction Δg_q, Equation (2.146) becomes:

$$\Delta F = n_q^\alpha \Delta g_q + \gamma^{\alpha\beta} A. \tag{2.163}$$

2.4.4.2 Heterogeneous Nucleation

As an example that is often taken up in practice, heterogeneous nucleation at the solid–gas and solid–liquid interfaces will now be considered.

Let a case in which a lens-shaped nucleus is formed on a smooth interface be considered, as shown in Figure 2.19. Herein, the liquid or gas phase in contact with a solid phase is the α phase, the phase formed by the nucleation is δ phase, and for simplicity both p^α and p^s are denoted as p ($p^\alpha = p^s = p$). The area of the solid interface that is in contact with the lens-shaped nucleus, $A^{\delta s}$, is expressed as:

$$A^{\delta s} = \pi d^2. \tag{2.164}$$

The Helmholtz energy before the nucleation occurs, $F_\circ$, is:

$$F_\circ = \sum_{i=1}^{r} n_{i,\circ}^\alpha \mu_i - pV_\circ^\alpha + \sum_{i=1}^{r} n_{i,\circ}^s \mu_i - pV_\circ^s + \sum_{i=1}^{r} n_{i,\circ}^{\alpha s} \mu_i + \gamma^{\alpha s} A^{\delta s} + F_r^s, \tag{2.165}$$

where $n_{i,\circ}^{\alpha s}$ is the number of moles of the component i adsorbed in the area $A^{\delta s}$ of the αs interface and F_r^s is the surface free energy at the outside of the interface with area $A^{\delta s}$.

Next, when the lens-shaped nucleus shown in Figure 2.19 is formed at a constant temperature and pressure, the free energy of the system is:

$$F = \sum_{i=1}^{r} n_i^\alpha \mu_i - pV^\alpha + \sum_{i=1}^{r} n_i^s \mu_i - pV^s + \sum_{i=1}^{r} n_i^\delta \mu_i - p^\delta V^\delta + \sum_{i=1}^{r} n_i^{\alpha\delta} \mu_i$$

$$+ \gamma^{\alpha\delta} A^{\alpha\delta} + \sum_{i=1}^{r} n_i^{\delta s} \mu_i + \gamma^{\delta s} A^{\delta s} + F_r^s. \tag{2.166}$$

Herein, it is assumed that the nucleus δ is at metastable equilibrium with the α and s phases and that the pressure and composition in these phases do not change even after nucleation because the size of the nucleus is extremely small, compared to the matrix α and s phases. Therefore, the relation $\mu_i = \mu_i^\alpha = \mu_i^s$ holds. By subtracting Equation (2.165) from Equation (2.166), the following is finally obtained:

$$\Delta F_g = -\left(p^\delta - p\right)V^\delta + \gamma^{\alpha\delta} A^{\alpha\delta} + \gamma^{\delta s} A^{\delta s} - \gamma^{\alpha s} A^{\delta s}. \tag{2.167}$$

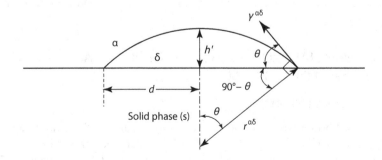

FIGURE 2.19   Lens-shaped nucleus on the flat interface between the α phase and solid phase.

From Young's Equation (2.84) and geometric relationship:

$$\gamma^{\alpha s} = \gamma^{\delta s} + \gamma^{\alpha \delta} \cos \theta = \gamma^{\delta s} + \gamma^{\alpha \delta} \left(\frac{r^{\alpha \delta} - h'}{r^{\alpha \delta}} \right); \qquad (2.168)$$

From Laplace's Equation (2.28):

$$p^{\delta} - p = \frac{2\gamma^{\alpha \delta}}{r^{\alpha \delta}}; \qquad (2.28')$$

and

$$V^{\delta} = \frac{2}{3} \pi r^{\alpha \delta^2} h' - 3\pi d^2 \left(r^{\alpha \delta} - h' \right). \qquad (2.169)$$

Substituting the above two equations into the first term of the right-hand side of Equation (2.167) and using the relation in Equation (2.168), the following is obtained:

$$\Delta F_g = \frac{1}{3} \gamma^{\alpha \delta} A^{\alpha \delta} + \gamma^{\delta s} A^{\delta s} + \gamma^{\alpha s} A^{\alpha s}. \qquad (2.170)$$

Herein, $A^{\alpha \delta} = 2\pi r^{\alpha \delta} h'$ and $A^{\alpha s} = \pi d^2$. In addition, $A^{\alpha \delta} = 2\pi r^{\alpha \delta^2} (1 - \cos \theta)$ and $A^{\delta s} = \pi r^{\alpha \delta^2} \sin^2 \theta$. Therefore:

$$\Delta F_g = \frac{2}{3} \gamma^{\delta} \pi r^{\alpha \delta^2} \left(1 - \cos \theta - \frac{1}{2} \cos \theta \sin^2 \theta \right). \qquad (2.171)$$

Meanwhile, $1 + \cos\theta + 1/2\cos\theta\sin^2\theta \geq 0$ or $1 \geq -\cos\theta - 1/2\cos\theta \sin\theta$. Therefore, ΔF_g in Equation (2.171) is:

$$\Delta F_g \leq \frac{4}{3} \gamma^{\alpha \delta} \pi r^{\alpha \delta^2} = \Delta F_{g,\text{sphere}}. \qquad (2.172)$$

The equal sign is applied to the condition for $\theta = 180°$, which is the case when the above equation is equivalent to Equation (2.149) for homogeneous nucleation. In other words, when $\theta < 180°$, the heterogeneous nucleation is more advantageous than the homogeneous nucleation in terms of free energy.

The above discussion on the interfacial phenomena in an equilibrium state, especially the description in Sections 2.2.2.3, 2.4.1, 2.4.3, and 2.4.4 is mostly based on reference 2. If you are interested and would like to understand it in detail, please refer to reference 2.

2.5 INTERFACIAL PROPERTIES AND PHENOMENA AT NON-EQUILIBRIUM

2.5.1 INTERFACIAL PROPERTIES

In a non-equilibrium state (except for the steady state), the chemical composition in the system changes with time. Therefore, the interfacial properties in the system, such as surface tension, interfacial tension, and wettability, also change. Strictly speaking, the relation between these interfacial properties and chemical compositions at non-equilibrium should be different from those at equilibrium. However, in case the chemical composition in the bulk near the interface can be estimated somehow, approximately estimating and understanding the interfacial properties at non-equilibrium using the relation between the interfacial properties and the above-mentioned chemical composition at or near equilibrium have been attempted. The examples include the analyses of wettability, penetration, and the Marangoni effect, which are discussed later. The adsorption of a surface-active component to the liquid phase interface is generally fast. Especially in the system of high-temperature melts described in Chapters 3 and 4, the above-mentioned approximation is not very far-fetched because the rate of adsorption becomes extremely high.*

Conversely, the phenomena in which the interfacial properties at non-equilibrium considerably deviate from the relation between the interfacial property and composition at equilibrium have been found. In what follows, several examples of such phenomena are introduced.

2.5.1.1 Surface Tension

As is described later (Section 3.1.1), in general, oxygen is known to be a strong surface-active element for molten alloys. Oxygen is also easily bound to many metals, resulting in the formation of oxides. In many cases, the equilibrium dissociation pressure of these oxides, $P_{O_{2,e}}$, is very small even at a temperature higher than the melting point of metals.

The sessile drop method is a widely used method for measuring the surface tension at high temperature in static states, i.e., the static surface tension. In this method, the surface tension is obtained from the measurement of the distortion of the droplet placed under gravity using Laplace's Equation (2.61). To control the partial pressure of oxygen p_{O_2} in the gas phase around the droplet, an inert gas such as Ar or He is flown around the droplet, but it is technically difficult to control the partial pressure p_{O_2} in the gas phase below $P_{O_{2,e}}$. However, in real experiments, it has been found for some systems that the surface tension measured under the condition where p_{O2} is considerably higher than $P_{O_{2,e}}$ (in bulk gas phase) is close to that measured

---

* When the speed of surface development (expansion) and extinction (contraction) is faster than that of the adsorption of the surfactant, the surface tension measured in a dynamic state, i.e., the dynamic surface tension, is different from the surface tension measured at the equilibrium state for the same composition material, i.e., the bulk phase. For example, the oscillating drop method, which is one of the methods used to measure the dynamic surface tension, is frequently used to measure the surface tension of the molten alloy.[17] In this method, both the expanding and contracting regions are generated at the surface with the oscillation of the droplet. In the expanding region, when the adsorption of the surface-active component is slower than the expanding speed of the surface, the surface tension of this region increases. On the other hand, if the desorption in the contracting region is slow, the surface tension decreases. In such a case, because the Gibbs–Marangoni effect is generated with the expansion and contraction of the droplet, it becomes necessary to reexamine the applicability of the calculation equation of the surface tension, which is derived by assuming that the surface tension of the droplet is uniform. The Gibbs–Marangoni effect is a change in the adsorption state of the surface-active component associated with the expansion and contraction of the surface causes the surface tension gradient, which leads to the Marangoni effect mentioned in Section 2.3.3. Further, the elastic behavior appears. This phenomenon can be referred to as the Gibbs–Marangoni effect.

at $p_{O_2} < p_{O_{2,e}}$. According to Passerone et al.,[18] the cause of this phenomenon can be interpreted by the extreme reduction in p_{O_2} near the droplet due to the reaction between the metallic vapor evaporated from the droplet and O_2 in the gas phase, as shown in Figure 2.20.

2.5.1.2 Interfacial Tension

Strictly speaking, the surface tension measured in the thermodynamic equilibrium state must be used in the analysis of the interfacial phenomenon in an equilibrium state, described in Section 2.4. Meanwhile, concerning the interfacial tension at the liquid–liquid interface, the composition of a phase cannot be changed independently from that of the other phase when the system is in an equilibrium state. For example, for a molten slag-alloy system, the composition of the slag changes with changes in the composition of the metal at a constant temperature and pressure. However, many of the previous measurement results have been reported in terms of the relation between interfacial tension and metal composition. Therefore, in terms of thermodynamics, many of these results should be treated as values measured in a non-equilibrium state.

When strong surface-active elements such as sulfur and oxygen are in a non-equilibrium state between the two phases, they are transferred across the slag–metal interface. It has been reported that when interfacial tension is measured by the sessile drop method, an extreme reduction in interfacial tension occurs during such a mass transport process (see the example in Figure 2.21[19]). It is considered that this phenomenon occurs because the interfacial state at the time of sulfur and oxygen transportation is different from that at equilibrium. Specific examples to explain it include the formation of an interfacially active intermediate compound during a reaction process and the accumulation of an interfacially active component at the interface in excess of the equilibrium amount.[20,21] Phenomenologically, it is proposed that the mass transport can be connected to the interface flow based on the thermodynamics of an irreversible process.[20]

The phenomenon of interfacial tension reduction probably needs to be discussed from another perspective. The sessile droplet method should be applied for conditions where Laplace's equation (Equation (2.61)), i.e., the static mechanical equilibrium, holds. Under the conditions where sulfur and oxygen migrate across the interface, it is fully conceivable that the concentration of the interfacially active component at the droplet interface becomes uneven. In this case, it is anticipated that

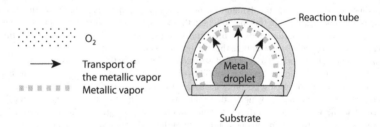

FIGURE 2.20 Antioxidant protective layer of the metallic vapor around the metal droplet.

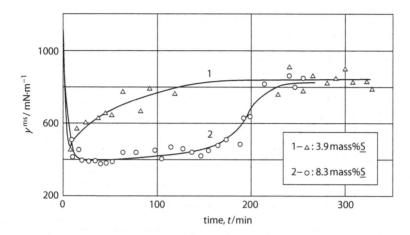

FIGURE 2.21 Time dependency of interfacial tension between the Fe–S alloy and CaO (40 mass%)–SiO (40 mass%)–Al$_2$O$_3$ (20 mass%) slag (1540–1560°C). (Adapted from reference 19.)

the droplet is distorted due to the generation of a strong Marangoni convection (see Figure 2.10 and Table 2.1 shown later). If the degree of this distortion is sufficiently large, the "reduction phenomenon" of interfacial tension obtained using Laplace's Equation (2.61) is a mere apparent phenomenon caused by the droplet distortion induced by Marangoni convection.

2.5.1.3   Wettability (Contact Angle)

《Time Dependency》

Let the time dependency of θ when the wetting between a droplet and a solid accompanies a reaction be considered. Figure 2.22 shows the time dependency of θ when molten 9.5at% Ti–Cu alloy is dropped on an Al$_2$O$_3$ substrate.[22] The stages I–III shown in Figure 2.22 can be interpreted as follows.[22] When the deformation of the droplet is considerably fast at the initial stage of the contact, stage I, which is where the mechanical equilibrium at the boundary of the droplet–ceramic–gas phases is not established, i.e., flow resistance, which is a rate-determining factor for the change in θ with time, shows up. When the rate of deformation of the droplet reduces, stage II shows up, where the changes in θ with time are determined by the reaction rate between the droplet and the substrate. In this case, the system is in a thermodynamic non-equilibrium state, but the three-phase boundary can be considered at mechanical equilibrium. At high temperature, since the reaction rate is generally high, in most cases, the mass transfer rate such as diffusion becomes the rate-determining factor of the reactions. Once it achieves thermodynamic equilibrium, the value of θ no longer changes with time and remains constant (stage III).

In addition, there is a report stating that, even for a metal droplet–ceramic system, the mass transfer between two phases sometimes affects the value of θ due to the reduction in γ^{ls}, similar to the case of interfacial tension. Further studies seem to be necessary on this matter from both experimental and theoretical viewpoints.

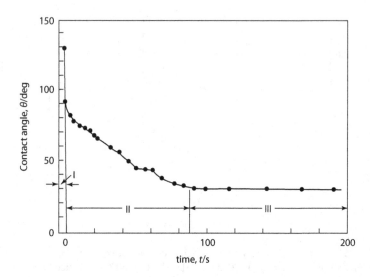

FIGURE 2.22 Time dependency of the contact angle between the Cu–9.5%Ti alloy and a ceramic substrate under vacuum condition at 1085°C. (Adapted from reference 22.)

When compounds are formed at a droplet–solid interface by a chemical reaction, the wettability between the compounds formed at the interface (formation of TiO in the case of the results shown in Figure 2.22) and the droplet should be considered as the wettability of the system.

When physical properties of the surface, such as the surface profile of the reaction product, changes because of certain formation reactions, for example, the surface of oxide becomes rough;[23,24] the reaction products cover a molten metal to form a shell;[25] the reaction products precipitate on the tip of a spreading surface and inhibit further spreading,[26,27] the influence of these aspects on θ needs to be considered.

《**Hysteresis**》

As shown in Figure 2.23, when the surface of a solid phase is tilted, it is often observed that the contact angle θ_a formed at the front, where the droplet slides off, increases, while the contact angle θ_r at the receding part decreases. θ_a and θ_r are called the advancing contact angle and receding contact angle, respectively. The contact angle can take an arbitrary value between the two extreme values of θ_a and θ_r depending on the condition. These properties are called wettability hysteresis. At thermodynamic equilibrium, once the state of a system is determined, there should be one corresponding θ value. Therefore, the wettability hysteresis phenomenon can be macroscopically or apparently classified as the wettability in a non-equilibrium state, as θ can take an arbitrary value depending on the conditions.

As to the hysteresis mechanisms, many theories have been proposed so far, including (1) friction theory,[28] (2) adsorption theory,[28] (3) rough surface theory,[28] and (4) surface heterogeneity theory.[23,29] It is considered that, in practice, the hysteresis of wettability should be explained as the phenomenon in which all these factors overlap.[28]

Considering that the surface and interface of solids, such as γ^s and γ^{ls}, are involved in wettability, another theory, (5) the surface stress theory, can also be further added. The surface stress of the solid surface increases due to being pulled at the three-phase boundary on the right-hand side of the droplet shown in Figure 2.23, which results in the compression of the solid–liquid interface. This situation enables the interfacial stress to decrease (see Section 2.2.2.6). Applying this result to Young's Equation (2.84), it is easily concluded that the contact angle θ decreases. The increase in θ can be explained in the same way as that for the three-phase boundary shown on the left-hand side of the droplet shown in Figure 2.23.

As a matter of a real problem, θ_a and θ_r themselves may be necessary in some cases. For example, θ_a may be necessary to investigate the penetration phenomenon (penetration rate) of molten steel or slag into oxide-based refractories and θ_r may be needed to analyze the initial state when non-metallic inclusions rise from the inside molten steel to the surface. Thus, when it comes to measurements, considering the measurement condition is necessary to obtain applicable results that can contribute to the specific application by considering the above factors.

2.5.2 INTERFACIAL PHENOMENON

Strictly speaking, most of the interfacial phenomena that actually occur and are observed are in a thermodynamic non-equilibrium state. In addition, there are many interfacial phenomena that are important and interesting from both scientific and practical (see Chapter 4) viewpoints. However, to understand these phenomena well and to quantitatively describe them, there are many things still unknown. Therefore, these phenomena have not yet been systematically arranged and described in contrast to the interfacial phenomena at equilibrium, as described in Section 2.4. Due to such a situation, the phenomenon tends to be explained individually, and the explanation tends to be from the engineering aspect. In this section, the interfacial phenomena in a non-equilibrium state will be discussed, and its fundamentals will be briefly explained.

2.5.2.1 Nucleation Rate

Needless to say, the nucleation rate is the interfacial phenomenon in a non-equilibrium state. It is important, together with the equilibrium theory (see Section 2.4.4) in

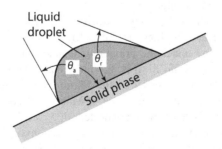

FIGURE 2.23 Hysteresis of wetting.

treating the nucleation that occurs in practice. Herein, the fundamentals of the homogeneous nucleation kinetics that have been often referenced are explained.

Let's consider a case in which a supersaturated component 1 in the β phase at temperature T creates I number of critical nuclei (α phase) per unit volume and unit time. Herein, it is supposed that the nucleation proceeds in a *quasi*-steady state and the critical nuclei are formed and grow by incorporating the embryos in the β phase. Then, they become the nuclei formation phase, and eventually the bulk phase. To describe these processes, the following equations, (2.173) and (2.174), are often used for condensed phase systems.[30,31]

$$I = A \exp\left(\frac{-\Delta F_g}{k_B T}\right), \tag{2.173}$$

$$A = n_g \left(\frac{\gamma^{\alpha\beta}}{k_B T}\right)^{1/2} \left(\frac{2 v_1^{\alpha,\circ'}}{9\pi}\right)^{1/3} n\left(\frac{k_B T}{h}\right), \tag{2.174}$$

where n_g is the number of molecules (or atoms) at the surface of the critical nucleus given by Equation (2.175), $v_1^{\alpha,\circ'}$ is the volume of a molecule (or an atom) of the nucleated α phase given by Equation (2.176), n is the number of molecules (or atoms) of component 1 per mole of matrix β phase, k_B is the Boltzmann constant, and h is the Planck constant.

$$n_g = \frac{4\pi r_g^2}{\left(\dfrac{M_1}{\rho_1^{\alpha,\circ} N_A}\right)^{2/3}}, \tag{2.175}$$

$$v_1^{\alpha,\circ'} = \frac{M_1}{\rho_1^{\alpha,\circ} N_A}, \tag{2.176}$$

where M_1 is the molecular weight of component 1, $\rho_1^{\alpha,\circ}$ is the density of component 1 in the α phase, and N_A is the Avogadro constant.

2.5.2.2 Marangoni Effect

The Marangoni effect described in Section 2.3.3 is a typical interfacial phenomenon observed in a non-equilibrium state. It might be better to term it anew as the "dynamic interfacial phenomenon."

The main characteristic of Marangoni convection is that its maximum flow rate is observed at the interface, as shown in Table 2.1,[32] and that, compared to the density convection occurring under the gravitational field, the smaller the volume is, or the thinner the liquid (liquid film) is, the more easily it occurs. This characteristic can be understood by the fact that the Marangoni number (equations (2.63) and (2.64)), as a measure of the occurrence of Marangoni convection, includes the term of the first power of the characteristic length L, while the Rayleigh number (Table 2.1), as a measure of the occurrence of density convection, contains the term L^3. Furthermore, the convex part of the interface is the suction point for Marangoni convection, whereas the convex part is the source point for density convection. It is

TABLE 2.1

Marangoni Convection and Density Convection[32]

Name	Marangoni Motion Marangoni Convection	Rayleigh Motion Density Convection
Common name	Interfacial disturbance	Natural convection
Driving force	Tension difference on an interface	Buoyancy (density difference)
Convex part at interface	Suction point	Spring point
Maximum flow rate	Interface	Inside
Sensitivity against contamination	Extremely sensitive	Almost no influence
Dimensionless number	Marangoni number Ma Equation (2.63), Equation (2.64)	Rayleigh number Ra $= g (\rho_i - \rho_0) L^3/D\eta$

ρ: density, L: the thickness of the liquid film, η: viscosity, D: diffusion coefficient, g: acceleration of gravity, suffix $_i$ indicates interface and suffix $_0$ indicates the value at the position of depth L.

generally difficult to distinguish Marangoni convection from density convection on the ground. However, if the shape of the interface (convex or concave) and the flow pattern (suction or source) can correspond to each other, it will become a very useful method to distinguish the two types of convection.*

Marangoni convection generally exhibits intense disturbance motion at the interface. This state is generally referred to as interface disturbance. The motion induced by the interface disturbance becomes most active near the interface, including the concentration boundary layer formed at the interface. Therefore, the mass transfer rate becomes extremely high. Since the reaction rate of a heterogeneous system containing high-temperature melt is generally determined by the mass transfer, Marangoni convection acts to increase the reaction rate of the heterogeneous system.[§] Specific examples can be found in gas–metal, slag–metal, and slag–metal-refractory reaction systems, which will be described in Section 4.4. Moreover, the influence of Marangoni convection on the reaction rate in the heterogeneous system is highly relevant to engineering, especially the practice of materials processing.

2.5.2.3 Dispersion

Generally, a system in which one phase is dispersed as a particle state in another phase is called a dispersion system. The particles are called dispersoid or dispersed phase, and the other phase (medium) is called the dispersion medium. The dispersion system is categorized into three classes according to the size of the dispersoid: (i) macroscopic (coarse) dispersion system, (ii) colloidal dispersion system, and (iii) molecular dispersion system. Since (iii) is a true solution system, herein only systems (i) and (ii) are dealt with. The difference between (i) and (ii) is not so clear; hence,

---

* Because density convection disappears in a microgravity environment, such as in an artificial satellite, research on Marangoni convection using such a microgravity environment has been actively conducted in recent years.[33]

§ However, the interfacial flow may be disturbed by the Marangoni effect, as described in Section 2.3.3. In this case, the mass transfer rate decreases.[13]

the two systems considered to be the same, for convenience. The dispersion system is also named after the type of dispersoid-medium combination: bubble for a gas–liquid combination, emulsion for a liquid–liquid combination, and suspension for a solid–liquid combination. In the dispersion system, interfacial tension exists between the dispersoid and the dispersion medium. Therefore, as far as the free energy of inside of the phase does not change due to the dispersion, the interfacial area, and thus the total free energy of the system, increase during the dispersion process. This implies that the dispersion system is in a thermodynamic non-equilibrium state (Figure 2.14).[16] To develop such a dispersion system, it is necessary to add energy larger than the energy as a driving force (driving energy), shown in Figure 2.24,[16] which considers kinetic factors in addition to ΔF^s, shown in Figure 2.14, which is the interfacial free energy increased by dispersion. As sources of this energy, the density difference between the dispersoid and dispersion medium, i.e., gravity, mechanical stirring, applied voltage, and interfacial tension gradient (see Section 4.2.3), which is proposed by the author, can be considered. Meanwhile, the dispersed particles tend to return to their initial state having a small interfacial area by aggregating each other. Taking Figure 2.24 as an example, the β phase particles are in a state in which they are naturally absorbed in the main body of the β phase. Therefore, when the dispersion system must be maintained for a long time, the "stability" of the system becomes an issue. The system becomes more stable with an increase in the driving energy (energy acting as the driving force) shown in Figure 2.24. The driving energy can be regarded as the potential energy of the so-called extinction process of the dispersed phase. The particle resistance force generated by the motion of particles, electric double layer, Saffman force (a kind of lift force), and van der Waals force can be considered sources of this energy.

《**Bubble**》

A gas dispersion system in a liquid phase is a bubble, which is more finely classified into bubble, foam, and dispersed gas. The bubble is an independent particle

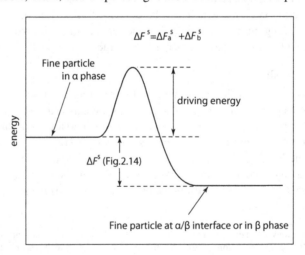

$$\Delta F^s = \Delta F_a^s + \Delta F_b^s$$

Fine particle in α phase

driving energy

energy

ΔF^s (Fig.2.14)

Fine particle at α/β interface or in β phase

FIGURE 2.24 Thermodynamic and kinetic illustration for the production and extinction processes in a dispersion system. (Adapted from reference 16.)

TABLE 2.2
Distinction Between the Foam and Dispersed Bubble[34]

	Foam	Dispersed Bubble
Explanation	State where bubbles gather and are separated by thin liquid or solid film.	State where many bubbles are dispersed in liquid or solid
Difference	Thin-film aggregation system	Bubble aggregation system
	Stability of the thin films dominates the stability of the system.	The viscosity of liquid and motion of bubbles dominate the stability
	The collapse of the thin films decreases stability.	The collapse of bubbles increases the degree of dispersion.
	Foaming agent	Blowing agent
	Defoamer (foam film collapse)	Defoamer (bubble rising and separation)
Example	Foam of soap	Pumice, Foam concrete, Turbidity of liquid due to bubble precipitation, Sponge, Soda pop, Soft serve ice cream
	Beer head	

comprising a gas phase existing in a liquid phase or surrounded by thin liquid films. The difference between dispersed gas and foam is shown in Table 2.2,[34] including the examples at room temperature, based on the references. The fundamental difference between the two is that each bubble in the dispersed gas system is independent and that the viscosity of the dispersion medium plays a major role in the transfer of the dispersed gas, while bubbles in the foam system become a polyhedron structure insulated by thin films, and the behavior of the bubbles is dominated by the property of the thin films. Therefore, when an external force is applied, for example, the form system becomes unstable due to the collapse of the thin films, while the dispersed gas system ends up being stable because of the micronization of the bubbles.

《**Emulsion and Suspension**》

The stability of emulsion and suspension is also an important issue in engineering. Since these dispersion systems are originally in a thermodynamic non-equilibrium state, the stability of the systems is equivalent to a kind of relaxation time required to achieve equilibrium state after the collapse of the dispersion system. The energy necessary for the collapse of the emulsion and suspension is expressed as the driving energy shown in Figure 2.24 in macroscopic perspective. From a microscopic perspective, one aspect of this destruction process is that it is induced by the fusion of particles. Another aspect is that the collapse is induced by the drainage, i.e., the process in which two liquid phases attempt to gather in one continuous phase by excluding one another. The fusion occurs in two stages, i.e., agglomeration and coalescence. Agglomeration is a phenomenon in which the droplets attach (through an interfacial film) by the attraction force between the droplets when two droplets come close due to the Brownian motion or motion caused by gravity. The same phenomenon can be found in suspension. Coalescence is the phenomenon in which the interfacial film between droplets collapses, and two droplets merge into one. This phenomenon is similar to the fusion of particles in suspension. Creaming is a phenomenon in which the drainage occurs prior to coalescence. Creaming is a process in which the droplets formed in the emulsion rise up or sink owing to the difference in the specific weight of the droplets and the dispersion medium. The separation state after this process is also referred to as creaming. The precipitation and floating of suspensions are similar to the creaming phenomenon. To control the emulsion, it is necessary to fully understand the factors promoting or inhibiting these processes. Such factors cited from the literature are summarized in Table 2.3,[35] focusing on the research mainly conducted at room temperature.

Thus far, the fundamentals of dispersion have been explained by focusing on the research results mainly conducted at room temperature. In Chapter 4, more specific examples will briefly be described. Nowadays, in the steel refining process, for example, argon gas and powder injection into molten steel, are conducted in various ways. Therefore, the handling of the dispersion systems composed of the metal–slag–gas phases caused thereby is also of great practical importance. In addition, the deoxidation process for molten steel, especially the process of formation and removal of non-metallic inclusions by forced deoxidation, involves nucleation and subsequent agglomeration, coalescence, and separation of fine particles. Therefore, the dispersion phenomenon is deeply involved in these processes. Furthermore, the foaming phenomenon of slag is also an important issue in engineering.

TABLE 2.3
Various Conditions Affecting Creaming, Aggregation, and Coalescence of Emulsion Particles[35]

	Inhibitory Factor	Promoting Factor
Creaming	Increase in viscosity of dispersion medium	The density difference between the dispersion medium and particle
Aggregation	Interaction of electric double layer *O/W* (interfacial potential, the thickness of double layer; the larger, the better) Ionic emulsifiers (+) Inorganic electrolyte (−) Increase in viscosity of dispersion medium	Brownian motion for long distance Fluid mechanical effect Temperature rise (+) Agitation (+) Gravity (+) Centrifugal force (+)
Coalescence	The steric effect of an adsorption film *W/O* (thickness of adsorption film, density; the larger, the better) Polymer material (+) Viscosity and elasticity of adsorption film Polymer material (+) Fine powder (−)	Dispersion force for short distance Adsorption film (−) Desorption of adsorption film High temperature (+) Addition of other components (+) Increase in interfacial tension

Note: *O/W*: Oil-in-water-type emulsion, *W/O*: Water-in-oil-type emulsion.
(+): positive effect; (−): negative effect.

The general expression about dispersion in this section is summarized based on references 34 and 35.

2.5.2.4 Penetration

As is apparent from Equation (2.86), a liquid with $\theta < 90°$* wets the solid–liquid interface and spontaneously penetrates a porous material, while that with $\theta > 90°$[18] does not penetrate spontaneously.

In the real penetration phenomenon, the process before the penetration depth reaches its maximum value (equilibrium state), i.e., penetration rate, often becomes an issue.

It is first experimentally verified that the penetration rate of water and organic liquid into porous materials such as filter paper and cellulose is given by the following Equation (2.177).[37]

$$l^n = kt,\tag{2.177}$$

where l is the penetration distance, k is a constant, and t is the time. For penetration into a single capillary (radius: r), $n = 2$, and thus:[38]

$$k = \gamma^1 \cos\theta \cdot \frac{r}{2\eta}.\tag{2.178}$$

In the case of vertical penetration under the gravitational field, the penetration rate dh/dt is:[39]

$$\frac{dh}{dt} = r^2 \frac{\dfrac{2\gamma^1 \cos\theta}{r} - \rho gh}{8\eta h}.\tag{2.179}$$

Since the static water pressure $-\rho gh$ is negligible at the initial stage of the penetration, the integration of the above equation produces the same type of formula as that in Equation (2.177), with $n = 2$, and where h is the penetration height.

For simplicity, a simply structured porous material composed of n capillaries with radii $r_1, r_2 \ldots r_n$ instead of complex porous structures of porous materials used in practice is now considered. The total volume V of the capillaries replaced by liquid penetration is:[40]

$$V = k\left(\frac{t}{\eta}\gamma^1 \cos\theta\right)^{1/2},\tag{2.180}$$

---

* For the penetration into a capillary having a uniform inner diameter or between flat plates having a constant clearance, $\theta < 90°$ is a condition under which the contact angle should be set such that immersional wetting occurs spontaneously. However, it has been reported that spontaneous penetration does not occur theoretically in a material wherein the spherical particles have the closest packed arrangement unless $\theta < 50.7°$.[36]

$$k = \frac{\pi}{\sqrt{2}} \sum_{i=1}^{n} r_i^{5/2}. \tag{2.181}$$

The pore structure of porous materials used in practice is complex; therefore, the practical penetration phenomenon cannot be fully explained by Equation (2.180).[40] To correct the complexity of the real porous material, the labyrinth factor ξ (also known as the tortuosity factor) is sometimes used.[41] However, the problem remains as to how ξ is estimated.

The penetration of high-temperature melts such as slag and metal into refractories is an important phenomenon in practice, which is deeply related to the lifetime of the refractories. However, since it is necessary to consider adsorption, reaction, etc., at the refractory–slag and refractory–metal interfaces for penetration in this system, the phenomenon becomes more complicated. Moreover, it is difficult to observe the phenomenon itself. In Section 4.1.1.2, examples of studies on such penetration phenomena of slag and metal into porous solid oxides (refractories) will briefly be introduced.

REFERENCES

1. E. A. Guggenheim: *Trans. Faraday Soc.*, **36** (1940), 397.
2. R. Defay, I. Prigogine, and A. Bellemans: *Surface Tension and Adsorption*, John-Wiley Sons, New York, NY (1966).
3. S. Ono: *Surface Tension*, Kyoritsu Shuppan (1980).
4. J. G. Kirkwood and I. Oppenheim: *Chem. Thermodyn.*, McGraw-Hill, (1961), 153.
5. S. Ono and S. Kondo: *Molecular Theory of Surface Tension of Liquid. Handbuch der Physik X*, Springer-Verlag (1960).
6. A. S. Skapski: *J. Chem. Phys.*, **16** (1948), 389.
7. C. Herring: *Structure and Properties of Solid Surfaces*, eds. R. Gomer and C. S. Smith, Univ. Chicago Press (1953).
8. K. Nii: *Bull. Jpn. Inst. Met.*, **13** (1974), 321.
9. R. Brückner: *Glastech. Ber.*, **53** (1980), 77.
10. J. Thomson: *Philos. Mag. Ser.* **4**, 10 (1855), 330.
11. C. G. M. Marangoni: *Ann. Phys.*, **143** (1871), 337.
12. F. D. Richardson: *Can. Metall. Q.*, **21** (1982), 111.
13. M. Sano and K. Mori: *Tetsu-to-Hagané*, **60** (1974), 1432.
14. C. V. Sternling and L. E. Scriven: *AIChE J.*, **5** (1959), 514.
15. M. Chikazawa and K. Tajima: *Kaimenkagaku*, Maruzen (2001), 32–33.
16. K. Mukai, T. Matsushita, and S. Seetharaman: Proceedings of Metal Separation Technologies III, Copper Mountain, CO, Helsinki Univ. of Technology (2004), 269.
    K. Mukai, T. Matsushita, and S. Seetharaman: *Scand. J. Metall.*, **34** (2005) 137–142.
17. J. Brillo, G. Lohofer, F. Schmidt-Hohagen, S. Schneider, and I. Egry: *Int. J. Mater. Prod. Technol.*, **26** (2006), 247.
18. A. Passerone, E. Ricci, and R. Sangiorgi: *J. Mater. Sci.*, **25** (1990), 4266.
19. A. A. Deryabin, S. I. Popel, and L. N. Saburov: *Izv. Akad. Nauk SSSR*, **5** (1968), 51.
20. A. A. Zhukhovitskii, V. A. Grigoryan, and E. Mikhalik: *Dokl. Akad. Nauk SSSR*, **155** (1964), 392.
21. K. Mukai, H. Furukawa, and T. Tsuchikawa: *Tetsu-to-Hagané*, **60** (1974), A7.
22. J. G. Li: *J. Mater. Sci. Lett.*, **11** (1992), 1551.
23. I. A. Askay, C. E. Hoge, and J. A. Pask: *J. Phys. Chem.*, **78** (1974), 1178.

24. K. Mukai, H. Sakao, and K. Sano: *J. Japan Inst. Met. Mater.*, **31** (1967), 923.
25. D. A. Weirauch: *Role of Interfaces*, eds. J. A. Pask and A. G. Evans, Plenum Press (1987), 329.
26. W. M. Armstrong, A. C. D. Chaklader and D. J. Rose: *Trans. Met. Soc. AIME*, **227** (1963), 1109.
27. A. P. Tomsia, E. Saiz, S. Foppiano, and R. M. Cannon: *High Temperature Capillarity*, Cracow, Poland (1997), 59.
28. The Chemical Society of Japan: *Encyclopedia of Experimental Chemistry*, Vol. 7, ed. K. Kagaku, Maruzen (1956), 72.
29. R. E. Johnson Jr. and R. H. Dettre: *Surface and Colloid Science*, Vol. 2, ed. E. Matijevic, Wiley-Interscience (1969).
30. D. Turnbull and J. C. Fisher: *J. Chem. Phys.*, **17** (1949), 71.
31. J. H. Hollomon and D. Turnbull: *Progress in Metal Physics*, Vol. 4, Pergamon Press, London, UK (1953), 342.
32. K. Fujinawa: *Trans. JSME* (in Japanese), **87** (1984), 1286.
33. T. Hibiya et al.: *Acta Astronaut.*, **48** (2001), 71.
34. T. Sasaki: *Kaimen genshou no kiso, Hyoumen kougaku kouza*, Vol. 3, ed. T. Sasaki et al., Asakura Publishing (1973), 155–202.
35. F. Kitahara: *Kaimen genshou no kiso, Hyoumen kougaku kouza*, Vol. 3, ed. T. Sasaki et al., Asakura Publishing (1973), 41–71.
36. G. Kaptay and D. M. Stefanescu: *AFS Trans.*, **213** (1992), 707.
37. J. M. Bell and F. K. Cameron: *J. Phys. Chem.*, **10** (1906), 658.
38. R. Lucas: *Koll. Z.*, **23** (1918), 15.
39. E. D. Washburn: *Phys. Rev.* **XVII** (1921), 273.
40. V. N. Eremenko and N. D. Lesnik: *The Role of Surface Phenomena in Metallurgy*, ed. V. N. Eremenko, Consultants Bureau Enterprises, Inc. (1963), 102.
41. Z. Li, K. Mukai, Z. Tao, T. Ouchi, I. Sasaka, and Y. Iizuka: *Taikabutsu*, **53** (2001), 577.

3 Interfacial Property of High-Temperature Melts

In this chapter, an overview of interfacial properties such as the surface tension of high-temperature melts, i.e., molten metal and molten slag, the interfacial tension between these two phases, and the wettability between the above-mentioned high-temperature melts and ceramics or refractories is presented.

The above-mentioned interfacial properties deepen our understanding of high-temperature melts, and they can be important fundamental data in solving technical issues in engineering. However, similar to the various types of measurements of the physicochemical properties of high-temperature melts, many difficulties exist in the measurement of the interfacial properties of high-temperature melts caused by phenomena such as reactions between the molten sample and the container or atmospheric gas and temperature control. In addition, in measuring the interfacial properties, control of the system is difficult because there are strong interfacial reactive components, which can easily cause contamination at the interface, resulting in further difficulty in the measurements.

First, the difficulties in making measurements are explained; thus, the points to be considered when using the measured values reported in the literature can be recognized. In particular, knowing the difficulties of making measurements is helpful for understanding the uncertainty of the measurements.

Next, an overview of the interfacial properties in high-temperature melt systems is presented, although the explanation is based on such measured values. Owing to space constraints our discussion is limited to a qualitative description, focusing on melts related to the iron and steelmaking process.

In the final section (Section 3.4), recommended data books and review papers are listed.

3.1 NOTES ON MEASUREMENT VALUES

3.1.1 MEASUREMENT ERROR[‡]

First, the measurement errors must be closely examined based on the reported information. The degree of accidental errors (dispersion of measurement values) and systematic errors (errors originating from the measurement principle or instrument itself, e.g., errors of the scale of a ruler) must be quantitatively evaluated as much as possible.

[‡] Translation supervisor note: With regard to the "error" in a measurement, it is recommended to assess and describe the uncertainty based on the Guide to the Expression of Uncertainty in Measurement (GUM) issued by the working group composed of several international organizations including the International Organization for Standardization (ISO).

Next, the conditions under which the values are measured must be confirmed and considered when choosing the most suitable values for your purpose. We hope that the previous chapter (Chapter 2) will help you judge this point. If possible, experimentally obtaining the values under the conditions consistent with the purpose of use and using these obtained values is preferable. The obtained values themselves will contribute to the elucidation of the interfacial properties of the high-temperature melts.

3.1.2 DIFFICULTIES IN MEASUREMENTS

3.1.2.1 Surface Tension of Metal

Let us discuss a measurement value of surface tension for pure molten iron (iron melt).

As shown in Figure 3.1,[1] the measurement values of the surface tension for iron melts tend to increase as the measurement date becomes more recent. This tendency is explained by the increased purity of iron melt over the years.[1] Furthermore, even if the measurements were conducted on dates close to each other, the measured values would differ from measurer to measurer. These facts show how large the systematic errors originating from the purity (composition) of the samples, and the measurement methods are, and also how large accidental errors are, due to factors such as various operational inaccuracies in the measurement (about ±3% in sessile drop method, which is often used in the measurement for molten metal).

In terms of the errors originating from the composition, particular attention must be paid to the concentration of strong surface-active elements, such as oxygen and sulfur. Figure 3.2[2] shows the measurement results of the surface tension for the Fe–O system. From a qualitative viewpoint, it is apparent that the surface tension

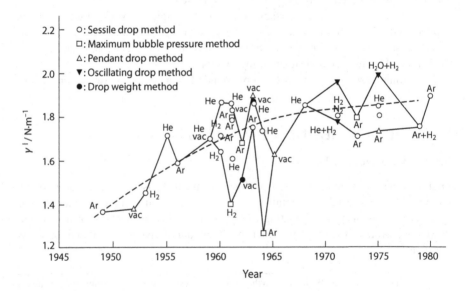

FIGURE 3.1 Measurement values and the measured date of the surface tension around the melting point of molten iron. (Adapted from reference 1.)

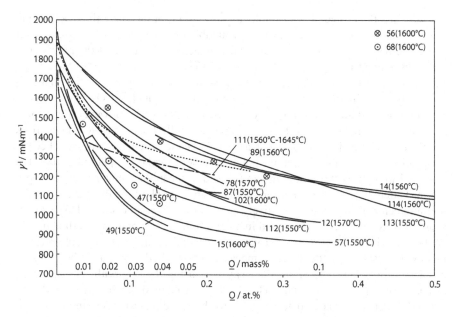

FIGURE 3.2 Measurement results of the surface tension of the Fe–O melt. In the figure, the number shown on the left side of the temperature corresponds to the reference number in the review article by Keene.[2] These numbers represent the values measured by several researchers. (Adapted from reference 2.)

strongly depends on the oxygen concentration. However, from a quantitative perspective, the differences among the measurers are also significant.

Let us now consider how to select the appropriate values suitable for the purpose of use from the measured values shown in Figure 3.2.[2]

First, it must be confirmed whether the measurement method is static (one of the representative methods is the sessile drop method) or dynamic (one of the representative methods is an oscillating drop method). Then, the measurement apparatus and methods must be evaluated; i.e., errors must be evaluated based on the measurement principle (method), and it must be confirmed whether the following information is clearly presented: (1) the reactivity between liquid metal and container or atmospheric gas, (2) the analysis results of the sample composition before and after the measurements, and (3) the reasonable measurement procedure along the principle. In particular, it is desirable for the partial pressure of oxygen in the atmosphere to be measured *in situ* (currently, it is relatively easy to measure this partial pressure using a solid electrolyte sensor).

3.1.2.2 Surface Tension of Slag

For molten slag, as in the case of steelmaking slag, strong surface-active components such as those seen in molten metal have not yet been found. Therefore, the errors originating from the sample composition are not so large as compared to those of the molten ferrous alloy. For the measurement of the surface tension of molten slag, the sessile drop method, and the maximum bubble pressure method are widely used. In the sessile drop method, there is a limitation in the selection of the material that supports the droplet. For example, for slag composition where the reaction with graphite can be ignored, graphite is most suitable for the substrate because the contact angle between the droplet and graphite is generally large and the precision of the measurement is high. In contrast, for slags containing easily reducible components such as iron oxide, the measurement is difficult owing to both the change in the composition by the reaction and the generation of bubbles (CO). In this case, a platinum or platinum–rhodium alloy is used as a substrate. However, the contact angle between the platinum alloy and slag is generally lower than 90°. Thus, if the sessile drop method is applied to such a system, the measurement errors will become extremely large.[3] In the maximum bubble pressure method, a platinum alloy that has good wettability with the slag can be used as a capillary material according to the measurement principle, and the method can be applied for the surface tension measurements of slags with a relatively large variety of compositions. However, the cause of systematic errors[4] observed in a maximum bubble pressure method has not yet been clarified. In terms of measurement principle, a maximum bubble pressure method is classified as the dynamic measurement method. Thus, a method must be developed that is capable of satisfactorily measuring the slag with a large variety of compositions in a static state. The detachment method has been reported[5] as a method which is considered to be close to the static state measurement method. However, this method has not yet been frequently used owing to the complexity of the apparatus and the systematic errors. The development of other methods in a static state has been reported,[6–8] but they have not been used by many measurers thus far.[‡]

3.1.2.3 Interfacial Tension between Slag and Metal

In the measurement of the interfacial tension between the slag and metal, the difficulties in measuring the surface tension of metal, i.e., the difficulty in controlling the surface-active components in the metal, and slag, i.e., the difficulty in controlling the reaction between the slag and the substrate, essentially overlap.

The most commonly used method to measure interfacial tension is the so-called sessile drop method, where the slag and metal (droplet) are placed in a crucible, and the interfacial tension is obtained by taking a picture of the shape of the metal droplet in the slag using the X-ray transmission apparatus. However, there is generally no crucible whose reactivity (e.g., solubility) with both slag and metal is sufficiently small under a specific condition, with a few exceptions. Therefore, the reaction between the crucible material and both phases would more or less affect the measurement of the slag–metal interfacial tension. It is, therefore, important to choose the most suitable material for the crucible.

It can be understood by actually taking pictures of the droplet shape that there are limitations on the transmission properties (resolution) of X-rays. Thus, only the blurred images of the metal droplet are obtained compared to those taken by the optical observation in the surface tension measurement. Thereby, the error in measuring the shape of the metal droplet becomes very large.

[‡] Translation supervisor note: A method to measure the surface tension and surface dilatational viscosity was suggested by the author (K. Mukai, T. Matsushita and S. Seetharaman, A novel experimental method for simultaneous measurements of surface tension and surface dilatational viscosity of liquid, CAMP-ISIJ, 23 (2010) 157).

To avoid reaction with the crucible material, another method, where a slag droplet is contacted to a large metal droplet, and the interfacial tension is then measured starting from the non-equilibrium state up to the equilibrium state by measuring its contact angle, has been reported.[9,10] However, since it is necessary to separately know the surface tension of the slag and metal in this method, the errors in these values must be considered. For this reason, this method has not yet been commonly used.

The important notices in treating the measured values of slag–metal interfacial tension have already been described in Section 2.5.1.2. It would be necessary to use the data with a good understanding of the meaning of the measured values as well as distinguishing between an equilibrium state and non-equilibrium state.

3.1.2.4 Wettability (Contact Angle)

As a quantitative measurement of wettability, we have already introduced the contact angle θ and changes in the interfacial free energies, w_s, w_i, and w_a, for wetting in Section 2.4.2.2.

The contact angle θ is often used as an intuitive measure of wettability. However, in most cases, the wetting between a high-temperature melt and a solid accompanies a reaction to a certain degree. In addition, there are not many solid phases with a flat surface. Furthermore, attention must be paid to the hysteresis of wetting. With respect to the above cases concerning the contact angle, see Section 2.5.1.3. Therefore, specifically for the measurement values of the contact angle, it is important to carefully confirm under what conditions the values are measured and try to choose the most appropriate values suitable for the particular purpose.

The contact angle θ is generally obtained by either of the following methods: (1) The value of θ is obtained by drawing a tangential line along the droplet surface as close as possible to the gas–liquid–solid three-phase boundary shown in Figure 2.13 and (2) the shape of the droplet is first obtained through computation based on Laplace's equation, and then the value of θ is obtained at the position where the lower part of the droplet is in contact with the solid substrate. Although method 1 has been used for a long time, it inevitably involves subjective errors. In method 2, since the shape is obtained based on many measurement points on the droplet surface, the problem in method 1 can be avoided. In reality, however, it is technologically difficult to clearly take a picture of the horizontal line on the substrate surface. Therefore, the obtained value of θ changes depending on the position of the horizontal line. In particular, the error increases as the value of θ approaches 180°.

As described above, the measurement of interfacial properties involves various difficulties. However, many researchers have challenged such difficulties associated with the measurement, and valuable data have been accumulated. In this section, rather too much focus was given on the problems and insufficiency of the measurement. However, when using the data, we would like you to boldly use these valuable results in both practical and academic aspects in accordance with the intended purpose, even if not sufficient. However, the results obtained using the measured values should be carefully interpreted.

3.2 SURFACE–INTERFACIAL TENSION

3.2.1 SURFACE TENSION OF METAL

The surface tension of metal is generally the largest among various liquids such as slag, salt, water, and organic liquids, as listed in Table 3.1.[11] The surface tension of slag and molten salt is smaller than that of molten metal, but larger than that of water and organic liquid. As shown in equation (2.43), the surface tension includes the term u^s, which corresponds to the internal energy. Moreover, the contribution of this term to surface tension is large (described later in this section). The internal energy of the liquid is closely related to the binding energy. It can be considered that the magnitude of surface tension in the various liquids described above reflects the strength of the bonding mode of the liquid, such as metallic bond (molten metal), covalent–ionic bonds (molten slag, salt), and van der Waals bond (molecular liquid).

As shown in Figure 2.4 in Section 2.2.2.4, the surface tension of various liquid metals extrapolated to absolute zero degree is linearly related to the heat of vaporization and the slope of this linear relation is slightly lower than 1/4. It was found that

TABLE 3.1
Surface Tension of Various Materials[11]

Materials	Surface Tension, γ^l/N m^{-1}	Temperature, T/K
Metal		
Ni	1.615 in He	1743
Fe	1.560 in He	1823
Ca	0.600	773
Covalent bond		
FeO	0.584	1673
Al$_2$O$_3$	0.580	2323
Cu$_2$S	0.410 in Ar	1403
Slag		
MnO·SiO$_2$	0.415	1843
CaO·SiO$_2$	0.400	1843
Na$_2$O·SiO$_2$	0.284	1673
Ionic bond		
Li$_2$SO$_4$	0.220	1133
CaCl$_2$	0.145 in Ar	1073
CuCl	0.092 in Ar	723
Molecular substance		
H$_2$O	0.076	273
S	0.056	393
P$_4$O$_6$	0.037	307
CCl$_4$	0.029	273

the surface tension of various metals at the melting point is also linearly related to the heat of vaporization and the slope of the linear relation is 0.17, i.e., also smaller than 1/4 (0.25).[12] However, when the value of $\gamma_m - T_m$ ($d\gamma/dT$) ($=u^s$, see equation (2.43)) is plotted against $(\Delta h_{m,\,vap} - RT_m)/(v_m^{2/3} \cdot N_A^{1/3})$ at the respective melting points, the result shown in Figure 3.3 is obtained (the suffix m represents the value at the melting point). The slope of the relation is 0.22, which is closer to 1/4 than the above values. The result is in good agreement with the fact that the structure of the liquid metal is supposed to be close to the face-centered cubic structure, i.e., the relation $\gamma_{mol} \fallingdotseq (1/4)\Delta h_{vap}$ in equation (2.46).

Concerning the surface tension of alloys with more than two components, many measurements have been performed. However, to the best of our knowledge, there have been no databooks that systematically and comprehensively present the measured results. If you need such data, start with reviewing the databooks and review papers listed in Section 3.4.

In this section, the surface tension of the binary alloys is only briefly presented in Table 3.2,[2] which summarizes the behavior of surface tension of binary iron alloys at low solute concentration. Table 3.2 shows that oxygen and sulfur are very strong surface-active elements. Furthermore, nitrogen and even alloying components such as Sb, Se, Te, Sn, and Y, are strong surface-active elements compared to various solute components in slag and aqueous solution.

For iron alloys alone, the values of the surface tension of a multicomponent low-concentration alloy using iron as a solvent is important in the steel refining process. However, the method of estimating the surface tension of such a dilute

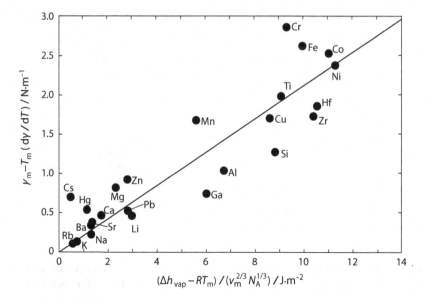

FIGURE 3.3 Relation between $\gamma_m - T_m(d\gamma/dT)$ and $(\Delta h_{vap} - RT_m)/(v^{2/3}N_A^{1/3})$ for liquid metal. The suffix m represents the value at melting point. (Adapted from reference 17.)

TABLE 3.2
Surface Activity of Solute *i* in Molten Iron

Solute (i)	Mean Value of Approximate Surface Activities (mN·m⁻¹·[at%i]⁻¹)	Approximate Range (in at%) over which Surface Activities was Derived
Al	-18	0–10
Sb	-2200	0–0.1
As	-540	0–1
B	-25	0–10
C	-4	0–10
Ce	0 or	
	-1750	0–0.04
Cr	-7	0–10
Co	-2	0–10
Cu	-30	0–10
Ga	-30	0–10
Ge	-55	0–5
La	0 or	
	-1500	0–0.4
Mn	-50	0–5
Mo	5	0–10
Ni	-2	0–20
N	-1400	0–0.1
O	-7490	0–0.03
P	-14	0–1
Pd	-17	0–5
Pt	-5	0–10
Rh	0	0–20
Se	-35000	0–0.01
Si	-13	0–5
S	-6310	0–0.05
Te	-190000	0–0.0025
Sn	-1630	0–0.15
Ti	0 or	
	-230	0–1
W	0	0–5
V	$+4$	0–5
Y	-2200 or	
	-6700 (with C present)	0–0.03
Zr	-1000	0–0.1

multicomponent solution that can withstand practical use has not been developed as yet.[‡] Therefore, at this moment, measurement must be conducted for each system.

[‡] Translation supervisor note: A paper was published by the author in 2008. K. Mukai, T. Matsushita, K. C. Mills, S. Seetharaman and T. Furuzono: Surface tension of Liquid Alloys – A Thermodynamic Approach, *Metal. Mater. Trans. B*, 39B (2008) 561–569.

Meanwhile, a method to estimate the activity coefficient of a solute component of multicomponent low-concentration alloys is already widely used. In this method, the activity coefficient is expressed as a function of the power of solute component concentration using interaction parameters.[13] This method may be a good reference for developing a method to estimate the surface tension of dilute multicomponent solutions.[‡]

The temperature coefficient of surface tension of a pure liquid is generally negative. In this case, the surface entropy s^s becomes positive, as apparent from equation (2.47). $s^s > 0$ indicates that the configuration of atoms (molecules) at the surface of a pure liquid is in a more disordered state than that within the liquid (see equation (2.4)).

The value of the entropy term Ts^s in equation (2.43) is estimated to be 700 mN/m for molten iron. In this case, the average value $d\gamma/dT \fallingdotseq -0.40$ mN/(m·K) of recently measured values summarized in Keene's review is used[2] as the temperature coefficient of the surface tension, and the temperature is considered to be 1,823 K. Similarly, the average value of the surface tension for molten iron at 1,823 K based on Keene's review is estimated[2] to be about 1,800 mN/m. Thus, the absolute value of the entropy term, 700 mN/m, corresponds to 40% of the value of surface tension. This shows that when the surface tension of molten iron (more generally, molten metal) is estimated, the entropy term must be considered.

The temperature coefficient of the surface tension of molten iron changes from negative to positive, with an increase in the concentration of oxygen or sulfur. This causes a large change in the penetration shape of the welding metal pool (discussed later in Section 4.2.1.1). However, for liquid silicon, the temperature coefficient of surface tension is negative up to the saturated concentration even if the oxygen concentration increases.[14,15] Based on this result, in the Czochralski method, which is used in single crystal growth for a silicon semiconductor, the change in flow direction along the temperature gradient of the liquid surface, which is found in the welding metal pool, cannot occur even if the oxygen concentration in the silicon melt increases.

3.2.2 Surface Tension of Slag

The surface tension of the major oxide components in the slag used for the iron and steelmaking process is about 0.3–0.6 N/m.[16] The change in the surface tension of the slag containing these oxides owing to change in composition is also small as compared with that of the molten metal. The following empirical equation (3.1) is generally used to estimate the surface tension of such a solution:

$$\gamma = \sum m_i J_i, \qquad (3.1)$$

where m_i is the mole% of component i and J_i is the so-called surface tension factor of component i. Boni and Derge[11] determined the surface tension factors, shown in

---

‡ Translation supervisor note: This method is taken in the abovementioned paper, K. Mukai et al., *Metal. Mater. Trans. B*, 39B (2008) 561–569.

Table 3.3, from the reported values measured for the 1–3-component oxide melts. In that study, using these values, the surface tension of the SiO_2–Al_2O_3–CaO–MgO–FeO five-component slag system was estimated and reported to be in good agreement with the experimental values.

Based on equation (3.1) and the surface tension factors listed in Table 3.3, it can be presumed that Al_2O_3, CaO, FeO, and MgO, which are the main components of slag, tend to increase the surface tension, whereas SiO_2 tends to decrease it. It can also be considered that B_2O_3, K_2O, and PbO are the components that significantly decrease the surface tension.

It is known that the surface tension of molten slag can also have a positive temperature coefficient depending on its composition. For example, the temperature coefficient of the surface tension of the binary slag has a positive value at high SiO_2 concentration,[11] and in the case of metasilicate slag, it changes from negative to positive with increasing ionic potential.[11]

TABLE 3.3

Surface Tension Factor of Oxides[11]

Oxide	Surface Tension Factor, J_i (mN·m⁻¹) 1300°C	1400°C	1500°C	Slag System	Concentration Range (mol%)
K_2O	168	156		K_2O–SiO_2	33–17
Na_2O	308	297		Na_2O–SiO_2	49–20
Li_2O	420	403		Lu_2O–SiO_2	46–29
BaO		366	366	BaO–SiO_2–Al_2O_3	50–34
PbO	140	140		PbO–SiO_2	83–33
PbO[a]	138[a]	140[a]		PbO	100
CaO		602	586	CaO–SiO_2	50–39
CaO		614	586	CaO–SiO_2	50–34
MnO		653	641	MnO–SiO_2	67–48
ZnO	550	540		ZnO–B_2O_3	67–52
FeO		570	560	FeO–SiO_2	77–60
FeO[a]		584[a]		FeO	100
MgO		512	502	MgO–SiO_2	51–46
ZrO_2		470[b]		Na_2O–SiO_2–ZrO_2	10–0
Al_2O_3		640[c]	630[c]	Al_2O_3	100
TiO_2		380		TiO_2–FeO	18–0
SiO_2[d]		285	286	Binary–SiO_2	83–50
SiO_2[e]		181	203	Binary–SiO_2	50–33
B_2O_3	33.6[a]	96[c]		B_2O_3	100

[a] Experimental value
[b] Value calculated by Dietzel
[c] Value extrapolated from the experimental value
[d] Calculated value from a binary system with high SiO_2 concentration
[e] Calculated value from a binary system with low SiO_2 concentration

---

3.2.3 SLAG–METAL INTERFACIAL TENSION

The interfacial tension between the slag and metal is generally close to the surface tension of the metal. The existence of strong interfacially active components, such as sulfur and oxygen, is similar to the case of the surface tension of the metal.

As described in Section 2.5.1.2, in strict terms, many of the measurements of slag–metal interfacial tension can be regarded as being conducted in the thermodynamically non-equilibrium state. Even with respect to the treatment of the measurement results, in most reports, the change in the interfacial tension is discussed by exclusively focusing on the change in the composition of the metal. Figure 3.4 summarizes the recent measurement results for various slag–molten iron interfacial tensions and oxygen concentrations in molten iron.[17] It can be seen that oxygen in the molten iron, especially in a low-concentration region, is a strong interfacially active element. Even if the composition of the slag is different, the interfacial tension is similar to a certain extent (about ±150 mN/m or less), thus implying that the concentration of oxygen in the molten iron is the dominating factor affecting interfacial tension.

Let us consider this result more specifically. When slag is in contact with molten iron and the system attains a thermodynamic equilibrium, the equilibrium of the reaction expressed by the following equation holds for oxygen.

$$\underline{O} = (FeO), \tag{3.2}$$

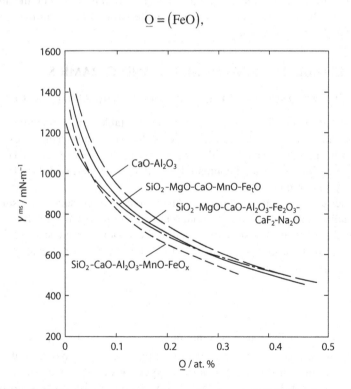

FIGURE 3.4 Relation between various slag–molten iron interfacial tensions and oxygen concentration in molten iron.

where $\underline{O}$ represents oxygen in the molten iron and (FeO) is the iron oxide in the slag. The equilibrium constant K_2 of reaction (3.2) is a function of only temperature and pressure, which is given by the following equation:

$$K_2 = a_{FeO} / a_O, \tag{3.3}$$

where a_{FeO} is the activity of FeO in the slag, and a_O is the activity of oxygen in the molten iron. Therefore, if the oxygen concentration in the molten iron is constant, the value of a_O can be regarded as nearly constant because the solubility of oxide in the molten iron is generally extremely small, and the concentration of the other solute components is low even if the composition of the slag is different. In this case, the value of a_{FeO} also becomes constant at constant temperature based on equation (3.3). However, if the slag composition is different, the FeO concentration in the respective slag is different because the activity coefficient of FeO is also different. Since the temperature of the system is high, the reaction rate is high. Therefore, it can be supposed that, at least at the slag–metal interface, reaction (3.2) is nearly at equilibrium, even if the system at the time of interfacial tension measurement is not strictly in equilibrium. In this case, from the results shown in Figure 3.4, it can be understood that the interfacial tension between the slag and Fe–O molten iron with almost the same a_{FeO} or a_O values (i.e., $\underline{O}$ concentration of the same degree) takes approximately the same value, even if the slag compositions are different. Based on the above discussion, it can be understood that oxygen in molten iron plays a dominant role in the slag–metal interfacial tension.

3.3 WETTABILITY BETWEEN METAL AND CERAMICS

3.3.1 CHARACTERISTICS OF WETTING BETWEEN MOLTEN METAL AND OXIDE

The wettability between the solid oxide, which is stable at high temperature and is used for refractory and molten metal, is generally bad. Considering an example for molten iron, the contact angle with oxides such as Al_2O_3 and MgO is $116°$–$135°$,[18] and the wettability is worse as compared with that of borides, such as TiB_2 and ZrB_2 ($0°$–$102°$[19]) and graphite ($50°$[20]).

Next, let us discuss the wetting for this system in more detail, focusing on the work of adhesion w_a of the oxide–molten metal system. It has already been found that the value of w_a is almost proportional to the standard free energy of formation $\Delta F_{i,f}^{\circ}$ of the oxide of molten metal i[21,22] (Figure 3.5[21]). The relation is given by the following equation:

$$w_a = w_0 + A_w \left(\Delta F_{i,f}^{\circ} \right), \tag{3.4}$$

where w_0 is the contribution of the binding energy based on the dispersion force, and $A_w (\Delta F_{i,f}^{\circ})$ corresponds to the contribution of the free energy based on the interaction and coordination states between the oxygen ions on the oxide surface and the metal. The stronger the affinity of metal with oxygen, the larger the contribution of the $A_w (\Delta F_{i,f}^{\circ})$ term. Based on the actual measured value of w_a (J/m²), the work of

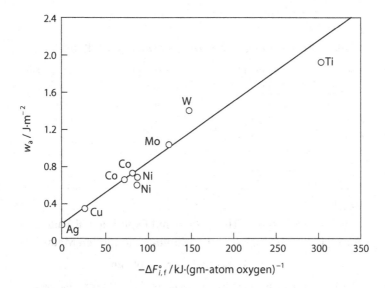

FIGURE 3.5 Relation between the work of adhesion between polycrystalline alumina and various metals, w_a (in a vacuum), and the standard free energy of formation of oxides $-\Delta F^{\circ}_{i,f}$. (Adapted from reference 21.)

adhesion for one mole of oxide surface $w_{a,mol}$ is calculated when the oxide surface is wetted with the molten metal. The results are summarized in Table 3.4[23] From these results, it can be interpreted that such a large $w_{a,mol}$ value cannot be generated by the van der Waals force; rather it can only be generated by the chemical interaction at the solid–liquid interface.[23] When the work of adhesion between pure metal and the $\langle 0001 \rangle$ plane of a single crystal of alumina is calculated from two separate parts, i.e., the work based on the chemical bonds between the metal atom and oxygen and the work based on the van der Waals force, the results shown in Table 3.5 are obtained.[21] The actual measured values agree well with the calculated ones. It can be understood that for Cr, Ti, and Zr, which have large $-\Delta F^{\circ}_{i,f}$ values, the contribution of the chemical bonds is very large as compared with that of the van der Waals force.

3.3.2 Effect of the Chemical Composition of Metal and Oxide

As apparent from equation (2.84), the contact angle θ is a function of γ^l, γ^s, and γ^{ls}. The values of γ^l and γ^s depend on the chemical composition of the liquid and solid phases, respectively. The value of γ^{ls} depends on the chemical composition of both solid and liquid phases. Therefore, it is easily understood that the contact angle θ depends on the chemical composition of both solid and liquid phases.

Figure 3.6[24,25] shows the example of changes in contact angle θ between the molten iron alloy and Al_2O_3 when the composition of the molten iron is changed by adding various components. As apparent from the figure, the addition of oxygen decreases θ most drastically. Then, the effect of addition on the decrease in θ becomes moderate

TABLE 3.4

Work of Adhesion $w_{a,mol}$ between Oxide and Molten Metal[23]

System	Temperature (°C)	w_a (10^{-3} J/m²)	$w_{a,mol}$ (kJ/mol)
Ni–ZrO$_2$	1500	917	59
Ni–CoO	1500	1500	71
Fe–Cr$_2$O$_3$	1550	1400	113
Fe–ThO$_2$	1550	1090	84
Cu–NiO	1100	990	42

TABLE 3.5

Comparison between Theoretical and Experimental Values for w_a between Single-Crystal Alumina and Molten Metal[21]

			Work of Adhesion w_a (10^{-3} J/m²)	
Metal	a	b	Theoretical Value (a + b)	Experimental Value
Ni	460	540	1000	1275
Cr	1150	590	1740	2020
Ti	1610	400	2010	2010
Zr	1950	365	2315	2320

a: Contribution of chemical bonding
b: Contribution of van der Waals forces

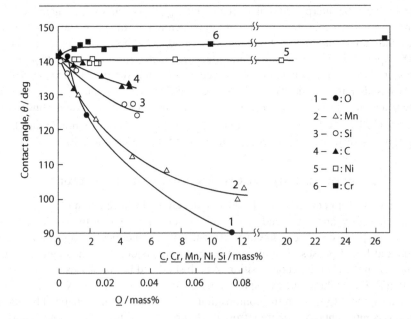

FIGURE 3.6 Effect of additive elements on the contact angle θ between polycrystalline alumina and molten iron (at 1500°C under Ar atmosphere). (Adapted from reference 24.)

in the order of Mn, Si, and C. The addition of Ni hardly changes θ and w_a, whereas the addition of Cr slightly increases θ. Furthermore, it has been recently reported that, concerning the relation between θ and the oxygen concentration in Figure 3.6, θ increases with the oxygen concentration in the region of low oxygen concentration (Figure 3.7).[26] The increase in θ associated with the oxygen concentration of up to around 100 ppm in Figure 3.7 is considered to be due to the decrease in Al concentration.

With respect to the changes in oxide composition, in the case of the oxide system containing components with low-solubility products in molten iron, such as the Al_2O_3–MgO system, θ hardly changes in the oxygen concentration range of 10–20 ppm in molten iron (low $\underline{O}$ concentration in the figure) even if the oxide composition changes, as shown in Figure 3.8.[26] However, in the system where the solubility products of oxides are large, θ decreases with an increase in Cr_2O_3 content, as shown in Figure 3.9.[27] This decrease is considered to be mainly caused by the increase in Cr concentration accompanied by the dissolution of Cr_2O_3 into the molten iron. Although there might be exceptions, the contact angle generally becomes lower, i.e., the wettability tends to become better, when the concentration of an element that is the same as the component in the ceramic compound is high, as shown in Figure 3.9. Let us expand this concept to the wettability of a metal with SiC, Si_3N_4, graphite, and diamond. The wettability between the above-mentioned ceramics and a metal with extremely low solubility of ceramics is bad.[28,29] However, the wettability between SiC and metals (iron, cobalt, or nickel), which can dissolve SiC fairly well, becomes better.[28,29] Similarly, the wettability between graphite and slag, which can hardly dissolve carbon, is bad. In contrast, the solubility of solid oxide in the slag is generally

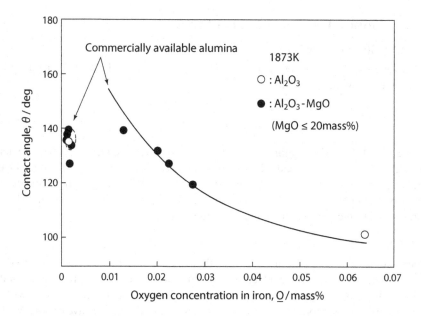

FIGURE 3.7 Relation between the contact angle of the Al_2O_3–MgO substrate and oxygen concentration in iron. (Adapted from reference 26.)

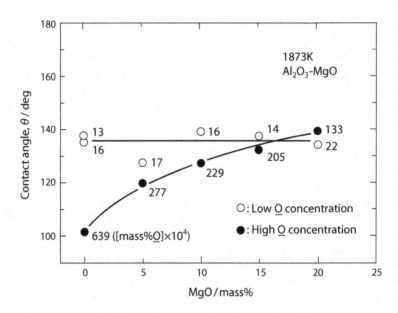

FIGURE 3.8 Relation between the contact angle of the Al_2O_3–MgO substrate–molten iron and the MgO content in the substrate. (Adapted from reference 26.)

high. In addition, the slag wets these solid oxides well. The expression "good wettability" here indicates that the contact angle is less than 90°, whereas "bad wettability" indicates that the contact angle is larger than 90°.

3.3.3 PHYSICAL FORM AND FACTOR OF SURFACE

3.3.3.1 Surface Roughness

For a rough surface, Young's equation (2.88) does not hold. Instead, Wenzel[30] proposed the following equation:

$$R_o\left(\gamma^s - \gamma^{ls}\right) = \gamma^l \cos\theta',\qquad(3.5)$$

where R_o ($=A/A_o$) is the roughness factor, A is the real surface area, A_o is the geometric area, and θ' is the so-called apparent contact angle against the rough surface. Figure 3.10[31] is an example where equation (3.5) holds. However, there are many cases that cannot be expressed by equation (3.5). Therefore, considering the shape and distribution of surface irregularities is necessary.

3.3.3.2 Structure of Interface

Microscopically, θ is influenced by the crystal orientation (Figure 3.11).[32] Macroscopically, when the contact angle is large (e.g., the contact angle in a metal-oxide system), the composite interface[33] shown in Figure 3.12, i.e., an interface with a structure where the molten metal cannot sufficiently enter the concave region, might be formed.

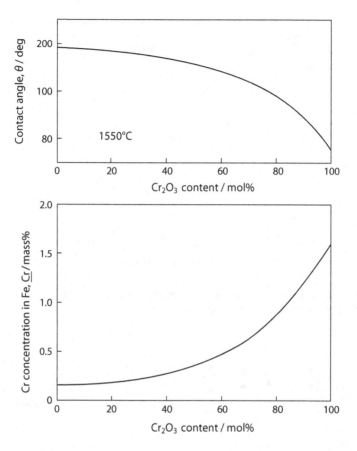

FIGURE 3.9   Influence of the Cr_2O_3 content on the contact angle of the Al_2O_3–Cr_2O_3 substrate. (Adapted from reference 27.)

In this case, Wenzel's equation is not enough, and the research conducted by Cassie and Baxter[34,35] will be helpful.

The (apparent) contact angle changes when the form of the surface changes, e.g., a ridge is formed at the tip of the spreading region by the metal-ceramics reaction as described in Section 2.5.1.3.

3.4   DATABOOK AND REVIEW PAPER

In this section, databooks and review papers published thus far, concerning the surface tension of the molten metal and the molten slag, the interfacial tension between these two phases, and the wettability between the above-mentioned melts and the ceramics, are chronologically introduced based on the author's best knowledge and their importance. However, publications that are extremely difficult to obtain owing to reasons such as being out-of-print or publications that are old but important data contained in them are expected to be included in new publications are omitted.

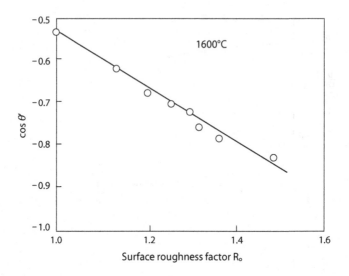

FIGURE 3.10   Relation between cos θ' and R_s for a molten iron–Al$_2$O$_3$ system at 1600°C (Wenzel's relation). (Adapted from reference 31.)

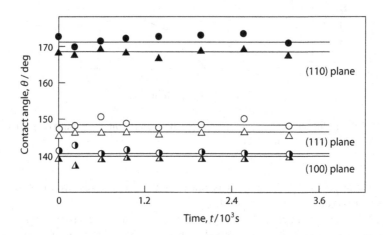

FIGURE 3.11   Time dependency of the contact angle between molten pure Sn and the low-index plane of the MgO single crystal under a hydrogen atmosphere at 600°C. (Adapted from reference 32.)

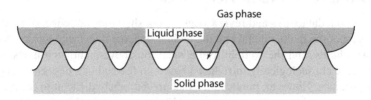

FIGURE 3.12   Composite interface at a rough surface. (Adapted from reference 32.)

3.4.1 DATABOOK

i) *Handbook of Physical Properties of Molten Iron and Slag*, Joint Society on Iron and Steel Basic Research, The Iron and Steel Institute of Japan (1972). In Chapter 4, data on the surface tension of molten alloy and slag, and the interfacial tension between these two phases are presented (by Y. Kawai and the other four authors).

ii) *Handbook of Physico-chemical Properties at High Temperatures*, The 140th Committee, JSPS, The Iron and Steel Institute of Japan (1988). In Chapter 5, data on the surface tension and wettability of metals and slag are presented (by K. Ogino).

iii) *Verein Deutscher Eisenhüttenleute: Slag Atlas*, 2nd Edition, Verlag Stahleisen GmbH, (1995). The data on the surface tension of slag are in Chapter 10; the data on the interfacial tension between slag and ferrous melts are in Chapter 11; and data on the wettability between ferrous melts and non-metallic solids are in Chapter 12 (by B. J. Keene).

iv) Y. Ishii, M. Koishi, and T. Tsunoda ed., *Wettability Technology Handbook—Fundamentals·Measurement valuation Data—Techno System* (2001). In Chapter 3, Part 2, the data on the wettability of ceramics materials are presented (by K. Nogi).

v) *Physical and Chemical DataBook for Iron- and Steelmaking—Ironmaking*, The Iron and Steel Institute of Japan, The 54th Committee on Ironmaking, Japan Society for Promotion of Science (2006). The surface tension and interfacial tension are presented in Section 28, and the data on the wettability are presented in Section 29 of Chapter IX (Editor chair: S. Inaba).

3.4.2 REVIEW

i) Ju. V. Naidich: The wettability of solids by liquid metals, *Progress in Surface and Membrane Science*, **14** (1981), 353.

ii) B. J. Keene: Review of data for surface tension of iron and its binary alloys, *International Materials Reviews*, **33** (1988), 1.

iii) B. J. Keene: Review of data for surface tension of pure metals, *International Materials Reviews*, **38** (1993), 157.

REFERENCES

1. T. Iida and R. I. L. Guthrie: *The Physical Properties of Liquid Metals*, Clarendon Press, Oxford, UK (1988).
2. B. J. Keene: *Int. Mater. Rev.*, **33** (1988), 1.
3. I. Jimbo and A. W. Cramb: *ISIJ Int.*, **32** (1992), 26.
4. K. Gunji and T. Dan: *Trans. ISIJ*, **14** (1974), 162.
5. T. B. King: *J. Soc. Glass Technol.*, **35** (1951), 241.
6. K. Mukai, H. Furukawa, and T. Tsuchikawa: *Tetsu-to-Hagané*, **63** (1977), 1484.
7. K. Mukai and T. Ishikawa: *J. Japan Inst. Met. Mater.*, **45** (1981), 147.
8. Z. Yu and K. Mukai: *J. Japan Inst. Met. Mater.*, **59** (1995), 806.
9. K. Mukai, H. Furukawa, and T. Tsuchikawa: *Tetsu-to-Hagané*, **60** (1974), A7.

10. J. L. Bretonnet, L.-D. Lucas, and M. Olette: *C. R. Akad. Sci. Paris Ser. C*, **280** (1975), 1169.
11. R. E. Boni and D. Derge: *Trans. Met. Soc. AIME*, **206** (1956), 53.
12. T. Tanaka, K. Hack, T. Iida, and S. Hara: *Z. Metallkd.*, **87** (1996), 380.
13. C. Wagner: *Thermodynamics of Alloys*, Addison Wesley Pub. Co, Inc. (1952).
14. Z. Niu, K. Mukai, Y. Shiraishi, T. Hibiya, K. Kakimoto, and M. Koyama: *J. Japan. Assoc. Crystal Growth*, **24** (1997), 369.
15. K. Mukai, Z. Yuan, K. Nogi, and T. Hibiya: *ISIJ Int.*, **40** (2000), Supplement, S148.
16. N. Ikemiya, J. Umemoto, S. Hara, and K. Ogino: *ISIJ Int.*, **33** (1993), 156.
17. K. Ogino: *Handbook of Physico-chemical Properties at High Temperatures*, eds. Y. Kawai, and Y. Shiraishi, ISIJ (1988), 170.
18. K. Ogino, A. Adachi, and K. Nogi: *Tetsu-to-Hagané*, **56** (1970), s 451.
19. G. V. Samsonov, A. D. Panasyuk, and M. S. Borovikova: *Sov. Powder Metall. Met. Ceram.*, **12** (1973), 476.
20. Yu. V. Naidich and G. A. Kolensnichenko: *Surface Phenomena in Metallurgical Processes*, ed. A. I. Belyaev, Consultants Bureau Enterprises, Inc. (1965), 218.
21. J. E. McDonald and J. G. Eberhart: *Trans. Met. Soc. AIME*, **233** (1965), 512.
22. S. Nakano and M. Ohtani: *J. Japan Inst. Met. Mater.*, **34** (1970), 562.
23. V. N. Eremenko: *The Role of Surface Phenomena in Metallurgy*, ed. V. N. Eremenko, Consultants Bureau Enterprises, Inc. (1963), 1.
24. B. V. Tsarevskii and S. I. Popel: *Izv. Vyssh. Ucheb. Zaved., Chern. Met.*, **8** (1960), 15.
25. B. V. Tsarevskii and S. I Popel: *Izv. Vyssh. Ucheb. Zaved., Chern. Met.*, **12** (1960), 12.
26. N. Shinozaki, N. Echida, K. Mukai, Y. Takahashi, and Y. Tanaka: *Tetsu-to-Hagané*, **80** (1994), 748.
27. K. Aratani and Y. Tamai: *Yogyo-Kyokai-Shi*, **89** (1989), 480.
28. K. Nogi and K. Ogino: *Int. Symp. Adv. Meter.*, Tokyo (1988).
29. K. Nogi: *Wettability Technology Hand Book—Fundamentals Measurement Valuation Data*, ed. Toshio Ishii et al., Techno System (2001).
30. R. W. Wenzel: *Ind. Eng. Chem.*, **28** (1936), 988.
31. K. Ogino: *Taikabutsu Overseas*, **2** (1982), 80.
32. K. Nogi, M. Tsujimoto, K. Ogino, and N. Iwamoto: *Acta Metall. Mater.*, **40** (1992), 1045.
33. R. E. Johnson, Jr. and R. H. Dettre: *Surface and Colloid Science*, Vol. 2. ed. E. Matijevtic, Wiley-Interscience (1969).
34. A. B. D. Cassie and S. Baxter: *Trans. Faraday Soc.*, **40** (1944), 546.
35. S. Baxter and A. B. D. Cassie: *J. Textile Inst.*, **36** (1945), T 67.

4 Interfacial Phenomena of High-Temperature Melts and Materials Processing

The currently conducted material manufacturing processes at high temperatures, such as metal refining, production of silicon single crystal for IC substrate, welding, and glass manufacturing, are mainly composed of a heterogeneous system containing two or more phases, such as molten metal, slag, glass, salt, and refractory for the container. Since the interface always exists in the heterogeneous system, the above-mentioned material manufacturing processes should involve more or less interfacial phenomena.

To appropriately control, improve, or develop such a material manufacturing process, there will be no objection that it can be one of the important approaches to well understand the physicochemical characteristics of the above-mentioned heterogeneous systems and pursue an accurate description and prediction method. For example, the steel refining process historically succeeded in describing and controlling the heterogeneous systems consisting of bulk phases such as gas, metal, and slag, and their phases using thermodynamics, transport phenomena, and electromagnetics. To aim for further development in the future, an understanding of the interface, more specifically, clarifying the influence of the interfacial phenomena, would become an important subject. The importance of this subject has been pointed out previously. For example, although it is slightly older, our review[1] describes the general matters concerning the steel refining process and interfacial phenomena in detail. However, the degree of its relation has not been sufficiently clarified specifically and experimentally until recently. In this chapter, we mainly introduce the results of the research on the relation between interfacial phenomena and steel refining processes, which we have been studying since the early 1980s, focusing on the research results of interfacial physical chemistry. Especially, it is noteworthy that the involvement of the Marangoni effect in the refining process has recently become clearer. Therefore, in Section 4.1, we will first briefly explain the relation of various interfacial phenomena with the refining process, other than the Marangoni effect. And then, in Section 4.2, we will introduce the relation of the Marangoni effect to materials processing with detail.

To understand such interfacial phenomena, it would be effective as well as important to understand it by directly observing the phenomenon that is really occurring. Therefore, we decided to provide a video clip showing various interfacial phenomena obtained through *in situ* observation or their relation with processing among our previous studies to the public on the internet as an additional resource.[1]

You should definitely watch the movie since the description of the text in Chapter 4 is brief, and the details of phenomena, which are difficult to explain in the text, are included in the video clip. The parts where the video clip is provided are marked with[†].

4.1 INTERFACIAL PHENOMENA IN THE STEEL REFINING PROCESS

As already described in Chapter 3, the surface tension and interfacial tension of the melts dealt with in steel refining processes, such as molten steel and molten slag, are larger than that of, for example, water by 5–20 times. In addition, since solutes such as oxygen and sulfur that unavoidably exist in steel refining processes are strong surface-active components, the interfacial phenomena described in Chapter 2 tend to be actualized in refining processes.

In the steel refining process, argon gas and powder are injected into the molten steel in various ways. The dispersion system consisting of the metal–slag–gas three phases, that is, the interface-evolved world, is considered to appear all over the process. Furthermore, the process of formation and removal of non-metallic inclusions in forced deoxidation progresses through nucleation and subsequent coagulation, coalescence, and separation of fine particles; therefore, the interfacial phenomena itself is deeply involved. Furthermore, the formation and control of the slag film, gas film, etc. which are described later may become an important technical issue in the refining process in the future.

4.1.1 WETTING

Wetting is a well-known interfacial phenomenon, which has been discussed in relation to the steel refining process. It has been clarified that wettability is also closely related to the behavior of argon gas that is injected in the continuous casting process. In addition, it has been shown that the penetration of slag or metal into refractories in which the wettability plays an essential role can be directly observed by an X-ray transmission apparatus. In what follows, these results are briefly introduced.

4.1.1.1 Behavior of Injected Argon Gas in a Continuous Casting Process[2–4†]

A water-model experiment showed that when argon gas is injected from a porous refractory at the inner wall of the immersion nozzle, the so-called stripe-like gas curtain is formed between the inner wall and molten steel, as shown in Figure 4.1, and the flow of molten steel becomes unstable because the lower end of this gas curtain breaks irregularly. This means that the gas curtain is formed by the bad wettability between the refractory and the molten steel, and disappears when the wettability is good, and then stable bubbles of uniform size are generated. This result shows that the gas curtain has the effect of preventing deposition of inclusions in the molten steel to the inner wall surface, i.e., nozzle clogging. However, it is presumed that the gas curtain promotes the turbulence of the molten steel flow, entrapment of the mold

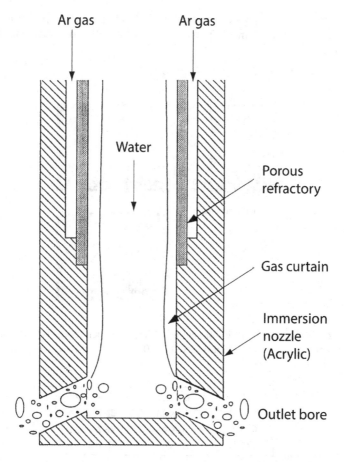

FIGURE 4.1 Schematic illustration of the gas curtain formed on the inner wall of the immersion nozzle under conditions of low wettability between the porous refractory and water (water-model experiment).

flux due to the formation of nonuniform bubbles, and entrapment of bubbles at the solidifying interface of molten steel.

4.1.1.2 Penetration of Slag[5] and Metal[6,7] into a Refractory[†]

《Penetration of Slag》

The penetration of slag into the refractory has a significant effect on the corrosion of the refractory because it causes the destruction of the bonds among the refractory constituent particles and the erosion or deterioration of constituent particles. Since the wettability between an oxide refractory and slag is generally good, penetration occurs spontaneously.

The *in situ* observation by the X-ray transmission apparatus revealed that the penetration of $CaO-SiO_2-FeO$ slag into a magnesia refractory is very fast. In the fastest case at the beginning of penetration, the penetration reaches a height of 20 mm in 10 s (Figure 4.2).[5] The penetration height h at the beginning of penetration is proportional to 1/2 power of time (Figure 4.3)[5] and formally coincides with Equation (2.177) at $n = 2$:

$$h = k_0 t^{1/2}, \tag{4.1}$$

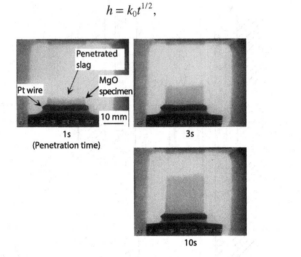

FIGURE 4.2 Penetration behavior of slag into MgO specimen (porosity: 31%) at 1,639 K. Composition of slag: mass% CaO/mass% $SiO_2 = 2.0$, T. Fe = 30 mass%. (Adapted from reference 79.)

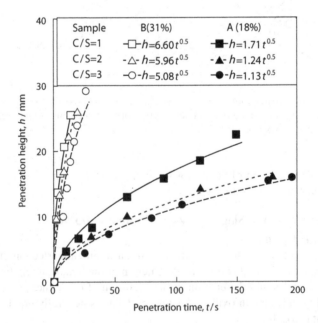

FIGURE 4.3 Temperature dependency of the penetration height of slag into MgO specimen at 1,693 K. A: porosity 18%, B: porosity 31%. Composition of slag: C/S = mass% CaO/mass% SiO_2, T. Fe = 30 mass%. (Adapted from reference 79.)

where k_0 is the constant determined by the experiment. The change in k_0 depending on the slag composition can be qualitatively explained using the physicochemical properties of this system included in k in Equation (2.178).

《**Penetration of Metal**》

Since the wettability between an oxide refractory for steel refining and molten steel is generally bad, the penetration of molten steel into the refractory does not occur spontaneously. However, for example, in the secondary refining of steel, since a porous plug refractory used for gas injection into molten steel in the ladle is installed at the bottom of the ladle, the penetration of molten steel occurs due to the metallostatic pressure of the molten steel. Such penetration is one of the causes of shortening the lifetime of the porous plug. We conducted an *in situ* observation of the penetration behavior of metal into the refractory for the porous plug when the metal phase side was pressurized using the experimental equipment incorporating the X-ray transmission apparatus (Figure 4.4). When the pressure of the metal side exceeds a certain value, the penetration begins from the portion having a large open pore diameter in the refractory. Unlike the case of slag (Figure 4.2), the height of metal penetration is different depending on the location; that is, it shows nonuniform penetration behavior (Figure 4.5). As the pressure is further increased, the rate of increase in the penetration height (equilibrium state) with respect to the increase in pressure becomes remarkably large above a certain

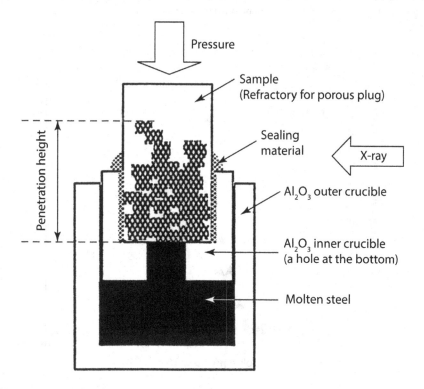

FIGURE 4.4 Schematic illustration of the experimental apparatus used for direct observation of the penetration of molten steel into the porous plug refractory.

value. In the case of mercury, the rate of increase in the penetration height is definitely proportional to the pressure (Figure 4.6).[6] It can be interpreted that the penetration at this stage is in a state where pores with a diameter that allows the penetration under the corresponding pressure are continuously linked over the entire refractory specimen. Besides mercury, the observation has also been conducted using silver and iron. For mercury, where the chemical reaction and adsorption of interfacially active

Applied pressure: (a) 0.13×10⁵ Pa, (b) 0.16×10⁵ Pa and (c) 0.24×10⁵ Pa

FIGURE 4.5 Penetration behavior of mercury into the porous plug refractory after the application of various pressures. (Adapted from reference 6.)

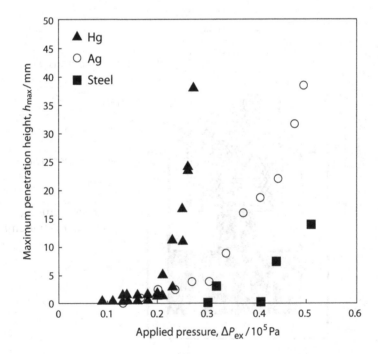

FIGURE 4.6 Relation between the penetration height of various molten metals into the porous plug refractory and the applied pressure. (Adapted from reference 6.)

components with refractory can be ignored, the behavior at the above-mentioned stage was clearly observed. In the case of silver, the chemical reaction can be ignored, but the behavior that is presumed to be the adsorption effect of dissolved oxygen in silver was observed. For iron, the chemical reaction with the refractory is supposed to occur besides the absorption reaction; therefore, the kinetic factor in the penetration behavior becomes apparent.

4.1.2 NUCLEATION OF ALUMINA IN ALUMINUM DEOXIDATION PROCESSES IN MOLTEN STEEL

In the deoxidation process of molten steel at the final stage of steel refining, aluminum is commonly used as a deoxidizer. However, there remain some unknown facts in the aluminum deoxidation process; for example, the origin of non-equilibrium phases of alumina, such as γ and δ phases,[8] which are observed at the initial stage of the deoxidation. Meanwhile, it has been reported that, when the deoxidation equilibrium between aluminum and oxygen is experimentally measured, a supersaturation phenomenon is found.[9] However, this phenomenon has also not been reasonably explained.

With respect to the origin of the non-equilibrium alumina phase produced from supersaturated molten iron, it is possible to predict the possibility of not only the production of the γ or δ phase of alumina but also the production of liquid alumina by combining the Ostwald's step rule and the classical nucleation theory (see Sections 2.4.4.1 and 2.5.2).[10] Such a process of non-equilibrium phase generation from a supersaturated state through nucleation can be considered as a phenomenon similar to the gas phase synthesis of diamond.[11]

The supersaturation phenomenon in aluminum deoxidation can be explained by adding the contribution of the change in free energy of the matrix phase produced accompanied by the nucleation reaction to the above-mentioned classical nucleation theory.[12] One form of the supersaturation phenomenon is that the system is in a state where the reaction cannot thermodynamically progress in the first place, i.e., the system is practically in the supersaturated state because the total free energy of the system is always increasing when the nucleation reaction is progressing. Another form of the supersaturation phenomenon can be explained as follows. The free energy of the system initially decreases as the nucleation reaction progresses but reaches a minimum when the nucleus grows to a certain size, unlike the case shown in Figure 2.18. Then, when the nucleus grows further, it reaches the state in which it cannot grow anymore because the free energy increases (see Figure 4.7[12]). The size of the nucleus corresponding to the minimum free energy state is said to be a few nanometers in diameter both theoretically[12] and experimentally.[13] When the fine Al_2O_3 particles of this size are suspended, an apparent supersaturated state emerges.

4.1.3 OTHERS

4.1.3.1 Dispersion

If it is possible to generate bubbles as fine as possible, when gas is injected from a porous refractory into molten steel, it is possible to increase the surface area between the gas and metal and improve the efficiency of the refining process.

Using the X-ray transmission apparatus, the bubbles formed on, and detached from, the surface of the porous plug refractory immersed in a Fe–C molten alloy are observed.[2] As a result, it was revealed that in the case of a Fe–C molten alloy, considerably large bubbles were generated as compared to those generated and detached when the same refractory was immersed in water (Figure 4.8). The fine bubbles generated from each pore on the refractory surface in the molten iron alloy coalesce with each other on the refractory surface to become one big bubble and detach. This

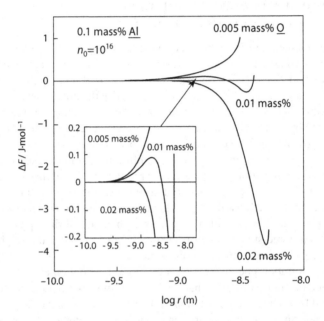

FIGURE 4.7   Free energy change when Al_2O_3 is nucleated in molten iron.[12] n_0: Number of nuclei in one mole of molten iron. (Adapted from reference 12.)

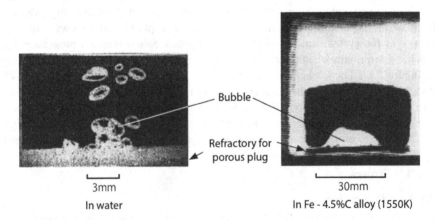

FIGURE 4.8   Size of bubbles generated from the porous plug refractory immersed in water and in a molten Fe–C alloy.

coalescence of fine bubbles results from the bad wettability between the refractory and molten iron alloy. From this result, it is understood that the generation of fine bubbles is difficult, in principle, due to the bad wettability, even when the pore size of the refractory for the porous plug is decreased.

Controlling the foaming of slag is one of the major technical issues in the steel refining process. It has been revealed from the direct observation, e.g., by the X-ray transmission apparatus, that the main cause of the foaming of slag is that micro-bubbles are generated at a high rate from the slag–metal interface due to the reaction between them.[14–16] It is thought that the generation of microbubbles is attributed to the good wettability between the slag and the metal and that the interfacial distur-bance caused by the Marangoni effect is involved in the generation of microbubbles at a high rate. Meanwhile, the experimental studies regarding bubbling into the slag have shown that the stability of the slag foams is dependent on the bubble size and that the foaming height is proportional to the gas supply rate (corresponding to the generation rate of CO bubbles).[17]

4.1.3.2 Adsorption

The nitrogen content in steel is closely related to the mechanical properties; there-fore, the control of nitrogen concentration in molten steel is important. For this rea-son, elucidating the reaction of molten steel with nitrogen in the gas phase, that is the reaction rate of nitrogen absorption and nitrogen desorption reactions, has drawn interest and studied extensively. The results showed that the existence of a small amount of a strong surface-active element, such as oxygen or sulfur, considerably reduces the reaction rate of the above-mentioned reactions involving nitrogen. For oxygen, one of the surface-active components, the surface is saturated with oxygen when, for example, its concentration in molten iron becomes higher than 200 ppm. In this case, it can be estimated that the surface oxygen concentration becomes higher than that in the bulk by more than 500 times. The reason why the reaction rate of nitrogen is extremely reduced is that the oxygen and sulfur adsorbed on the surface act as an interface resistance in terms of the reaction rate at the interface. This sce-nario is called the surface resistance model and is now generally accepted.[18,19]

However, most of the above experiments on reaction rate have been carried out under vigorous stirring, e.g., using an induction furnace or an oscillating droplet using the levitation technique.

Meanwhile, it has been revealed that when an electric resistance furnace which does not induce the inductive stirring is used, the nitrogen reaction rate can be rea-sonably explained by taking the contribution of the Marangoni convection caused by the concentration difference of one of the surface-active components, i.e., nitro-gen,[20,21] at the surface of the molten iron into account based on the direct observation and analysis of Marangoni convection at the surface of the molten iron.*[22–25]

---

* The influence of oxygen in molten iron can also be explained as follows. The decreasing rate of the sur-face tension of molten iron because of the concentration of nitrogen N is significantly decreased owing to the presence of O.[21] Therefore, even if the nitrogen concentration gradient at the surface of molten iron is the same, the surface tension gradient decreases owing to the presence of O. Consequently, the Marangoni convection is weakened, and the molten iron–nitrogen reaction rate becomes slow.

4.2 MARANGONI EFFECT IN MATERIALS PROCESSING

As described above, the surface (interfacial) tension of molten steel, molten steel–slag, and molten slag is extremely high compared to that of an aqueous solution. In addition, there are strong interfacially active components such as oxygen and sulfur in these systems. Therefore, even if the density or viscosity of molten steel and slag is somewhat high, there is a large possibility that the interface disturbance occurs at the interface of the melt due to the Marangoni effect when the interfacially active components transport across the interface (see Sections 2.3.3 and 2.5.2.2). Also, since a large temperature gradient is likely to occur at the surface (interface) of high-temperature melts, it is also highly likely that Marangoni convection is generated by the temperature gradient.

Meanwhile, in the system at high temperature and with entangled complex phe-nomena in the gravitational field on the ground represented by the metal refining process, it seems an undeniable doubt for many engineers and researchers as to "whether the Marangoni effect really occurs or not," even today. However, the gen-eration status of the Marangoni effect in high-temperature melts has been gradu-ally resolved in these 20 years or so. Along with that, considerable progress has also been made in the demonstrative clarification of engineering issues involving the Marangoni effect. In what follows, the generation status of the Marangoni effect in high-temperature melts and its relation with the materials processing, focusing on our research results are described in detail.

4.2.1 DIRECT OBSERVATION OF MARANGONI EFFECT OCCURRING IN HIGH-TEMPERATURE MELTS

4.2.1.1 Marangoni Convection Due to the Temperature Gradient
《Liquid Column of Molten Salt》[†]

The surface tension gradient caused by the temperature gradient expressed by Equation (2.62) is considered as the driving force of liquid motion under the condi-tion where we can assume that only the temperature gradient exists along the surface. However, in the gravitational field, the contribution of the density gradient caused by the temperature gradient also needs to be considered. The difficulty in verifying the Marangoni effect in the gravitational field is to separate the contribution of the density convection in the flow from that of Marangoni convection. We[26] used an apparatus* that applies the principle of the hot-thermocouple method[27] to directly observe the status of the flow in a liquid column by steadily giving the temperature gradient in a small liquid column, and we obtained the following results. The flow of molten $NaNO_3$ sandwiched by the upper and lower Pt plates (diameter: 2–3 mm) showed different specific flow patterns depending on the given temperature distribu-tion (Figure 4.9).[26] These flow patterns and flow rates can be investigated using the

---

* Using a V-shaped Pt–Pt·Rh wire pair welded to the upper and lower Pt plates (for heating and measur-ing the temperature), a temperature gradient is generated in the vertical direction on the free surface of the liquid column that is sandwiched between the two platinum plates (diameter: 2–3 mm). The surface tension gradient caused by the temperature gradient is thus obtained.

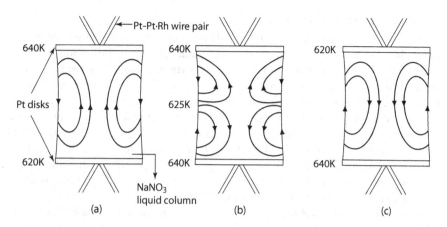

FIGURE 4.9 Marangoni convection generated in a NaNO₃ liquid column at various temperature distributions. (Adapted from reference 26.)

fine platinum particles suspended in the melt as markers. Figure 4.9a shows the case where the upper and lower ends of the liquid column are maintained at high and low temperatures, respectively. There is a very fast flow from the high-temperature region to the low-temperature region on the free surface, whereas, in the central part, there is a flow slower than that on the free surface from the low-temperature region to the high-temperature region. Since the temperature coefficient of the surface tension of molten $NaNO_3$ is negative, the lower is the temperature, the higher is the surface tension. Therefore, the direction of the observed flow agrees with the direction of flow induced by the Marangoni effect. Furthermore, when the temperature of the upper region is kept low, and that of the lower region is kept high, a flow pattern having a direction exactly opposite to that shown in Figure 4.9a is observed (shown in (c)). If density convection is dominant in the case of (c), a flow from high temperature (lower end) to low temperature (upper end) should be generated in the center of the cylinder. Figure 4.9b shows the case where the temperature of the central region is set at low. In this case, the direction of the flow is from the high-temperature region to the low-temperature region on the free surface, and four vortex flow patterns are observed from the observation position. In the case of molten NaOH with a positive temperature coefficient for surface tension, the flow pattern having a direction exactly opposite to that of $NaNO_3$ is observed. The flows observed in liquid columns (a) and (b) can almost be quantitatively explained to be caused by the Marangoni effect based on the numerical analysis results using the finite difference method.[28]

These results indicate that, if these melts have the surface tension gradient of about several hundred N/m², a strong flow that can be regarded as caused by the Marangoni effect is generated even in the gravitational field, and that such flow is extremely fast even in a small region, such as a liquid column of about 1 mm thickness. On the basis of the L-dependence of the Marangoni number (Ma) described in Section 2.3.3, Marangoni convection is generated preferentially to density convection in such a small liquid column, even in the gravitational field, where the liquid has a size of about ϕ 3 mm × 3 mm.

《Weld Pool During Welding》

Since the temperature gradient in a weld pool during welding is large, the generation of Marangoni convection due to the temperature gradient is fully anticipated.[29–31] According to the observations of laboratory experiments using iron, the penetration shape is shallow when the oxygen concentrations are low and deep when they are high.[32] These results correspond well with the fact that the temperature coefficient for surface tension changes from negative to positive at several tens of ppm as the concentration of oxygen increases (Figure 4.10).[33,34] This is because the surface flow from the central part of the weld pool (high-temperature region) to the peripheral part (low-temperature region, liquidus temperature) is generated when the oxygen concentration is low, while that from the peripheral part to the central part is generated when the oxygen concentration is high (Figure 4.11).[‡]

In practice, during welding, the penetration shape becomes shallow when the sulfur concentration is low and deep when the sulfur concentration is high, similar to the oxygen case. This result corresponds well with the fact that the temperature coefficient for surface tension changes from negative to positive as the sulfur concentration increases (Figure 4.10).[33,34] However, unlike the observations of laboratory experiments, in practical welding works there are difficulty in controlling oxygen partial pressure, the variety of metal composition, and the possibilities of the existence of an oxide layer on the weld pool surface. It is therefore necessary to consider these factors in the analysis.

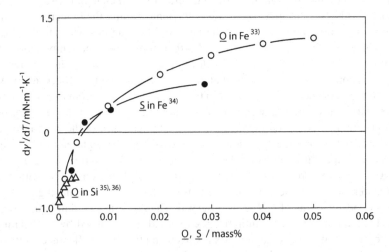

FIGURE 4.10 Changes in temperature coefficient for the surface tension of molten iron and silicon by oxygen and sulfur concentrations.

[‡] Translation supervisor note: Somewhat similar phenomena may be found in the additive manufacturing (AM) process as well. For example, in powder bed fusion (PBF) AM through selective laser melting (SLM), the powder layers are locally melted and the molten pool is formed. The convection in the molten pool may be induced by the Marangoni effect, as described here, and the shape of the molten pool, and the solidification behavior might be influenced by the Marangoni convection.

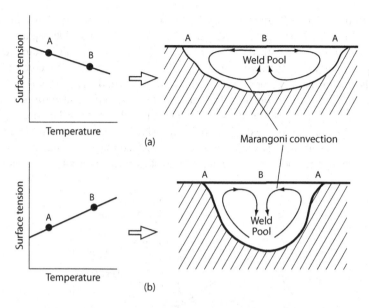

FIGURE 4.11 Marangoni convection in a weld pool and the shape of the weld pool. (a) When the concentration of the surface-active component is low. (b) When the concentration of the surface-active component is high.

⟪Silicon Liquid Column⟫

Hibiya et al. succeeded in detecting Marangoni convection in a silicon liquid column with the temperature gradient under microgravity using the X-ray transmission apparatus.[37] They further investigated, in detail, the relation between the status of Marangoni convection and the Marangoni number or partial pressure of oxygen in the gas phase by observing the temperature oscillation of the liquid column using a thermocouple.[37,38]

4.2.1.2 Expansion and Contraction of a Slag Droplet Caused by Electric Potential Change[†]

In a system with electrocapillarity, the driving force of liquid motion under the conditions where only the change in electric potential exists along the interface may be the interfacial tension gradient caused by the change in electric potential in Equation (2.62). The possibility of the existence of electrocapillarity at the slag–metal interface has been indicated.[39] However, in the slag–metal system, the composition of the melt also changes due to the electrochemical reaction associated with the electric potential change. Thus, it is possible that the change in surface tension due to the change in the electric potential is not completely separated from that due to the change in the composition in the previous studies. We found that when the surface of molten Pd is in contact with a PbO–SiO_2 slag droplet and the applied voltage ψ between the two phases is rapidly changed; a reversible expansion–contraction behavior is generated in the slag droplet.[40] More specifically, when a PbO–30 mol%SiO_2 slag droplet (0.1 g) on molten Pb is supported by a Pt–20%Rh wire under Ar atmosphere at 1,073 K, as shown in Figure 4.12, and when applied voltage between the Pb (WE) and Pt–Rh wire (CE) is rapidly changed, the slag droplet repeats the reversible expansion–contraction motion along with the change in the applied voltage. For example, when the applied voltage is changed from 0 to −2 V, the slag droplet first contracts, then expands, and eventually stops after 5 s. Thereafter, when the applied voltage returns again to 0 V, the slag droplet first contracts, then expands, and returns to almost the same state as the initial state at 0 V, and eventually, stops. When the angle α in Figure 4.12 is measured to calculate the metal–slag interfacial tension γ^{ms} under the mechanical equilibrium assumption using the surface tension values for metal and slag, the change in the γ^{ms} value becomes as shown in Figure 4.13.[40] Specifically,

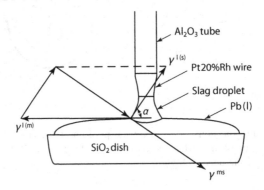

FIGURE 4.12 Shape of a PbO–SiO$_2$ slag droplet in contact with the molten Pb surface and the visible angle α.

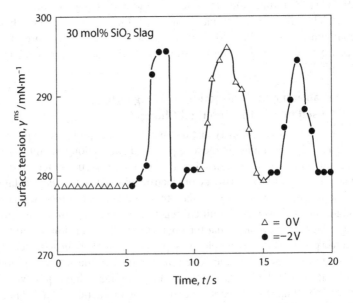

FIGURE 4.13 Time dependency of the (PbO–SiO$_2$ slag)–Pb(l) interfacial tension with changes in applied electric potential.

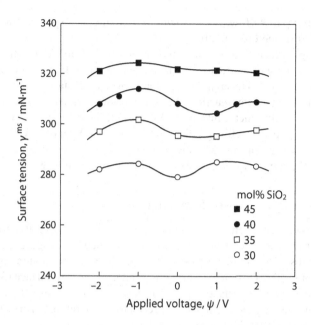

FIGURE 4.14 Influence of applied electric potential on the interfacial tension of the PbO–SiO$_2$ slag–Pb (l).

when ψ is changed from 0 to -2 V, γ^{ms} first changes to a maximum value and eventually settles to the same value at $\psi = -2$ V in Figure 4.14.[40] It can be considered that such changes in γ^{ms} occur along the γ^{ms}-ψ curve shown in Figure 4.14.[40] In addition, this change is almost reversible. It is difficult to think that the rapid expansion–contraction motion of the slag droplet and the reversible changes in γ^{ms} caused by the changes in the slag composition due to the slag–metal reaction. Rather, it is conceivable that the rapid motion is caused mainly by the changes in surface tension caused by the electrocapillarity, i.e., the Marangoni effect.

Following this, Toguri et al.[41] directly observed the motion of Cu$_2$S, FeS, matte, and copper droplets in slag caused by the electric potential difference and reported that the moving speed of the liquid droplets increases with the applied electric field and droplet diameter.

4.2.1.3 Motion of Slag Film Caused by the Concentration Gradient

The driving force of melt motion under the conditions where only the concentration gradient of the interfacially active component exists along the interface may be the interfacial tension gradient caused by the concentration gradient in Equation (2.62). Nonetheless, even in such a system, the contribution of the density gradient caused by the concentration gradient must be considered in the gravitational field, similar to the case described in Section 4.2.1.1. Moreover, it is quite difficult to verify the concentration gradient at the interface of high-temperature melts in a fluid state. The possibility of the generation of Marangoni convection caused by the concentration gradient has been indicated in some pyrometallurgical reaction systems in the past.* However, sufficiently convincing experimental results had not been obtained before our studies.

At the slag surface or at the slag–metal interface of an oxide refractory, a slag film is formed along the refractory surface because the wettability between the slag and refractory is good. Through the reaction between this slag film and refractory or slag film–refractory–metal reaction, the concentration gradient is formed on the film surface or at the film–metal interface. As a result, active Marangoni convection is generated by the concentration gradient in the slag film. Moreover, we found through a direct observation that this Marangoni convection is the main cause of the local corrosion of the refractory. This is explained in detail in Section 4.2.2.

4.2.2 Local Corrosion of Refractory

It is well known that a refractory for steel refining is often locally corroded (local corrosion) at the slag–metal interface or slag–gas interface. Figure 4.15 shows the corrosion status of the main blast furnace trough by slag and molten pig iron.[43] The occurrence of local corrosion is clearly observed near the slag surface (SL) and slag–molten pig iron (ML) interface. The local corrosion of the refractory is a significant problem affecting the lifetime of the refractory. Therefore, an effective measure against this problem has long been awaited.

4.2.2.1 Oxide Refractory

《**Previous Research**》

Historically, at first, local corrosion at the slag surface of the oxide refractory was taken up as a practically significant problem for the refractory of a tank for glass melting. The major achievements of research conducted in this field include, but are not limited to, Brückner,[44,45] Dunkel and Brückner,[46] and Busby's research[47] and reviews. In summary, the local corrosion at the slag surface is caused by the interfacial disturbance due to the Marangoni effect of slag, mainly near the solid-oxide–molten-slag–gas three-phase boundary. However, these studies were limited to model experiments conducted around room temperature, centered on the aqueous solution

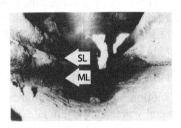

FIGURE 4.15 Local corrosion of the main blast furnace trough, as observed from the tap hole side. SL: slag line, ML: metal line.

---

* As shown in the author's review,[42] more than a dozen research results have reported the possibility of the generation of Marangoni convection and its influence on the reaction rate in gas–metal and slag–metal reactions.

system around and the qualitative estimation based on the observation and analysis of samples after the solidification for the real refractory–slag system. Therefore, it is difficult to say that these studies were sufficiently convincing. For example, 9 years after the study by Jebsen-Marwedel,[48] which is regarded as the first study that proposed the original concept of the mechanism, Vago and Smith[49] proposed a theory that the main cause of the local corrosion is the presence of volatile components in slag. Thereafter, Löffler,[50] who investigated the local corrosion phenomena in detail by mainly using solidified samples, also did not completely deny the contribution from the volatile components. In the 1980s, Caley et al.[51] proposed a theory that the local corrosion at the $PbO–SiO_2$ slag surface is mainly caused by oxygen in the gas phase. Also, concerning the local corrosion of oxide at the slag–metal interface, Brückner,[45] Sendt,[52] and Schulte[53] presumed that a local concentration difference was generated at the glass–metal interface around the oxide–molten-glass–metal three-phase boundary, similar to the slag surface, generating the difference in surface tension to cause the interfacial disturbance, which then became the main cause of the local corrosion. However, in these studies, the local corrosion phenomena were not fully grasped from both macroscopic and microscopic viewpoints. In addition, even the existence of the surface tension gradient, which is considered essential to verify the existence of interfacial disturbance due to the Marangoni effect and that of the liquid phase motion (Marangoni convection) near the three-phase boundary, was not confirmed. In 1979, Iguchi et al.[54] tried to explain the local corrosion of the oxide refractory from electrochemical mechanisms based on the fact that the local corrosion of Al_2O_3 at the molten metal–$PbO–SiO_2$ slag interface was influenced by the applied voltage. As seen above, with respect to the local corrosion in this system, a convincing answer to the questions, "What is the main cause of the local corrosion?"; "Is it caused by the combination of many factors?"; and "Or is it caused by a completely different mechanism?" had not yet been obtained. Meanwhile, it has recently been revealed that the Marangoni effect plays a major role in local corrosion. More specifically, we found, through a series of experiments and analysis based on the *in-situ* observation using an optical apparatus and X-ray transmission apparatus, that the local corrosion of the oxide refractory is induced by the mass transfer of soluble components from the refractory being effectively promoted and "breaking the diffusion layer," because the slag film or slag meniscus produced on the above-mentioned slag surface or the refractory surface near the slag–metal interface actively moves due to the Marangoni effect.

《Local Corrosion in the System Where the Surface (Interfacial) Tension Increases by the Dissolution of Refractory Components》

$(SiO_2(s)–(PbO–SiO_2)$ Slag System)[†]

In $SiO_2(s)–(PbO–SiO_2)$ slag systems[55-57] where the dissolved components increase the surface tension of the slag, the flow of the slag film on the vertically immersed cylindrical SiO_2 surface shows a flow pattern consisting of a wide upward flow zone and a narrow downward flow zone, as shown in Figure 4.16.[56] For cylindrical specimens (Figure 4.16a), both flow zones shift with time. In the case of a square column specimen (Figure 4.16b), the upward flow zone is always observed at the ridge, while the downward flow zone always observed in the flat region. Since SiO_2 dissolves as the slag film rises in the upward flow zone, the concentration of SiO_2 in the slag film becomes higher as the flow moves upward. In the case of $PbO–SiO_2$ slag, surface tension increases with the increasing of concentration of SiO_2.[58] Therefore, the film is pulled upward, and the continuous upward flow is maintained as shown in Figure 4.17. When the slag is accumulated in the upper region due to the upward flow, the film in this region becomes thicker. Since the surface tension of the upper film is high, this thickened film tends to form a droplet. This droplet-like part is eventually pulled downward by gravity and becomes a downward flow. The local corrosion of this system progresses in

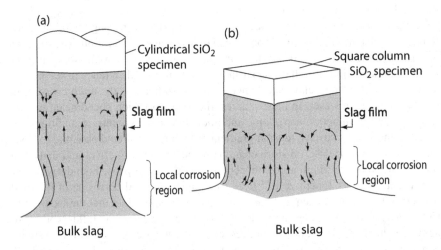

FIGURE 4.16 Flow pattern of the $PbO–SiO_2$ slag film generated on a SiO_2 specimen surface. (Adapted from reference 56.)

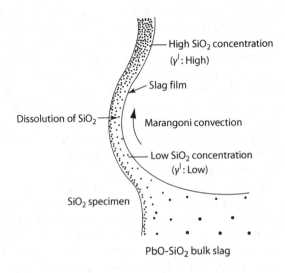

FIGURE 4.17 Mechanism of the upward flow (Marangoni flow) generation of the slag film shown in Figure 4.16.

the region where the specimen surface is actively washed, as described above, due to the fresh slag film supplied from the bulk. In particular, the upward flow zone of the slag film becomes the main part for promoting the local corrosion. In the case of a square column specimen, the ridge is always washed by the upward flow, as shown in Figure 4.16b. Therefore, the corrosion speed becomes extremely high compared to that in the flat region, where the surface is washed by the downward flow with high SiO_2 concentration. Eventually, the original square shape in the horizontal section in the local corrosion part becomes a near-round shape over time.

The flow of the slag film in the upward flow zone can be explained well based on the result of fluid dynamical analysis using the measured surface tension gradient and film thickness.

(SiO_2 (s)–(PbO–SiO_2) Slag–Pb (l) System)[†]

The local corrosion of SiO_2 (s) also occurs at the Pb–slag interface when Pb and PbO–SiO_2 are melted in a transparent SiO_2 crucible.[59,60] In this case, the slag film is formed between Pb (l) and the inner wall of the SiO_2 crucible, and the existence of the downward and upward flows is confirmed by direct optical observation. Since the dissolution of SiO_2 into the slag film increases the Pb–(PbO–SiO_2) interfacial tension,[40] the slag film is pulled down due to the Marangoni effect induced by the SiO_2 concentration gradient in the vertical direction of the slag film, and the downward flow is generated (Figure 4.18). This situation can be considered to be the case exactly when the motion of the slag film of the SiO_2 (s)–(PbO–SiO_2) slag system, shown in Figure 4.17, is turned upside down. In this case, the gas phase is replaced by Pb (l). Such slag film motion effectively promotes the mass transfer of the dissolved component (SiO_2) by basically the same mechanism of local corrosion in the above-mentioned SiO_2 (s)–(PbO–SiO_2) slag system, i.e., the system

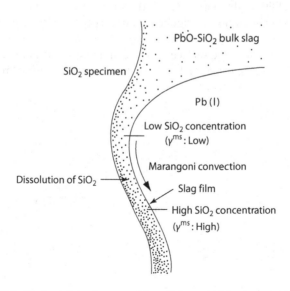

FIGURE 4.18 Mechanism of the downward flow (Marangoni flow) generation of a slag film in the region of local corrosion of the (PbO–SiO_2 slag)–Pb(l) interface.

where the dissolved component increases the surface tension of the slag induces the local corrosion.

Other examples of such local corrosion in the systems where the surface (interfacial) tension is increased, as described above, include the magnesia-chromium refractory–slag system[61] and blast furnace trough material–slag-molten pig iron system.[43,62,63]

《Local Corrosion in the System Where Surface (Interfacial) Tension Decreases by the Dissolution of Refractory Component》[†]

In the SiO_2 (s)–(Fe_tO–SiO_2) slag system,[64] where the dissolved component decreases the surface tension, the slag meniscus rotates along the surface of the cylindrical specimen in such a way that a transverse wave washes a quay or repeats upward and downward movements as a whole in such a way that a wave comes and goes to the beach, as shown in Figure 4.19.[65]

The factors considered to dominate such motion of meniscus include (i) Changes in the surface tension and density of the slag meniscus accompanied by the dissolution of SiO_2, (ii) changes in the contact angle between solid SiO_2 and slag, and (iii) the Marangoni effect. In this system, since the change in the contact angle is too small to observe, factor (i), i.e., the changes in the shape of the meniscus determined by the static balance expressed by Laplace's equation (2.61), are considered to have a dominant effect. However, to analyze the motion of the meniscus more accurately, the contribution of factor (iii) or factor (ii), depending on the type of the system, must be considered. In addition, it is experimentally demonstrated that the active Marangoni convection, in the direction of flow away from the specimen (Figure 4.20), is generated on the meniscus surface due to the formation of the concentration gradient with the dissolved component (SiO_2). For the system where the local corrosion occurred mainly by the meniscus motion, the constriction is sharp, and the vertical width is narrow. In addition, the horizontal section of the local corrosion for the square column specimen maintains its original square shape while the local corrosion progresses, unlike the SiO_2 (s)–(PbO–SiO_2) case.[65] Other examples of such local corrosion in the systems where the surface (interfacial) tension is decreased, as described above, include the trough material–slag system[66] and magnesia-chromium refractory–slag–molten steel system.[67]

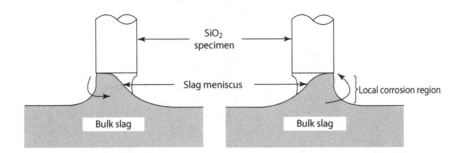

FIGURE 4.19 Motion of the FeO–SiO_2 slag meniscus generated on the SiO_2 specimen surface. (Adapted from reference 65.)

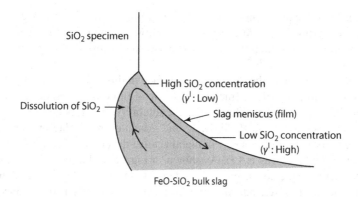

SiO₂ specimen

Dissolution of SiO₂

High SiO₂ concentration
(γ^l: Low)

Slag meniscus (film)

Low SiO₂ concentration
(γ^l: High)

FeO–SiO₂ bulk slag

FIGURE 4.20   Mechanism of the Marangoni flow generation of a slag meniscus in the region of the local corrosion of the SiO₂ (s)–(FeO–SiO₂) slag system. (Adapted from reference 65.)

《System Where the Surface Tension Does Not Change Even When the Refractory Component Is Dissolved》

Meanwhile, the surface tension of Na₂O–SiO₂ slag is almost constant even if SiO₂ concentration increases.[68] Although the slag film is formed along the specimen surface around the slag–gas interface, even when a solid SiO₂ specimen is partly immersed in Na₂O–SiO₂ slag, neither the slag film motion nor the local corrosion is observed depending on the experimental condition.[69]

4.2.2.2   Oxide–Non-Oxide Composite Refractory†

In recent years, there has been a huge change in the refractory material for steel refining, that is from conventional oxide refractory to composite refractory where non-oxides such as graphite are added to oxides. However, even for such a composite refractory, noticeable local corrosion occurs at the slag–metal (molten steel) or slag–gas interface, especially at the former.

Hauck and Pötschke[70] investigated, through immersion methods, the local corrosion at a powder–metal interface of an immersion nozzle material (main components: Al₂O₃ and graphite) for continuous casting using CaO–Al₂O₃–SiO₂ slag and CaO–Al₂O₃–CaF₂ slag. Slag was found between the metal and the nozzle material in the local corrosion region after solidification. They presumed that a chemical reaction between the nozzle material and slag induced a gradient in the nozzle material–slag–metal interfacial tension near the nozzle material–slag–metal three-phase boundary, which caused interfacial disturbance, which in turn, coupled with an agitation effect of CO gas, increased the mass transfer rate, which caused the local corrosion. To the best of our knowledge, this paper was the first one which is published in an academic journal in the field of metal refining to have clearly described that the Marangoni effect might be a major cause of local corrosion. However, as in the previous studies on local corrosion, this paper also lacked sufficient experimental proof of the phenomena from both macroscopic and microscopic viewpoints. Therefore, it was difficult to say whether the proposed mechanism of local corrosion was sufficient in terms of validity and concreteness.

We conducted experiments using a system similar to that of Hauck et al.[70] From a series of experiments and analyses of *in situ* observation using an X-ray transmission apparatus, we revealed that the local corrosion in this system progresses with the following mechanism.[71,72]

The local corrosion at the slag–metal interface in this system occurs by the cyclic up-and-down motion of the refractory–slag–metal three-phase boundary, as shown in Figure 4.21,[71] due to the difference in the solubility of components of the refractory (oxides, such as Al_2O_3,[71,72] MgO,[73,74] and graphite) in the slag or metal and the difference in the wettability of slag or metal against the components of the refractory (oxides and graphite) (Table 4.1). As shown in Figure 4.21a, in this stage, the slag penetrates between the nozzle material and metal to form the slag film, and the oxide in the nozzle is dissolved into the slag film. As a result, when the surface of the nozzle material becomes graphite-rich, the slag film, which has bad wettability with graphite, is repelled and recedes. Then, the surface of the nozzle material is wetted

TABLE 4.1

Solubility of Oxide and Carbon into Metal (Molten Steel) or Slag. Wettability between Oxide or Carbon and Metal (Molten Steel) or Slag.

	Solubility		Wettability	
	Slag	Metal	Slag	Metal
Oxide	High	Very low	Good	Bad
Carbon	Low	High	Bad	Good

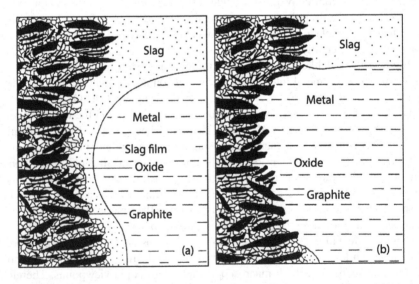

FIGURE 4.21 Mechanism of local corrosion at the slag–metal interface of an oxide–carbon refractory.

with metal, which has good wettability with graphite, and the slag–metal interface rises (Figure 4.21b). In this stage, graphite, which is directly in contact with metal, is rapidly dissolved into the metal. When the nozzle material surface consequently becomes oxide rich, the slag that has good wettability with oxide penetrates from the upper slag phase, and the slag film is formed again. The local corrosion progresses by the repetition of these processes. Therefore, the shorter is the time required for one cycle of the up-and-down motion of the slag–metal interface, the higher is the rate of local corrosion. In the stage where the three-phase boundary is located in the lower part (Figure 4.21a), the dissolution of oxide components into the film is promoted by the Marangoni convection of the slag film, similar to the local corrosion at the slag–metal interface in the oxide refractory. In the present system, bubbles are actively generated from near the refractory–slag–metal three-phase boundary at the local corrosion region. Especially in the case of magnesia-carbon refractory, the higher the graphite content in the refractory and the iron oxide concentration in the slag become, the more active is the bubble generation. Eventually, a gas layer, the so-called gas curtain, is formed between the refractory and the slag. This gas curtain acts as a protective layer for the refractory and decreases the rate of local corrosion.[74]

As described above, the nature of local corrosion of the refractory can be regarded as a typical interfacial phenomenon in which the Marangoni effect, wetting, and other parameters play important roles. Note that the "surprising surface tension effect" was revealed through the research on local corrosion. For example, in a system where the dissolved component of solid oxide increases the surface tension of the slag, such as the $SiO_2(s)$–$(PbO–SiO_2)$ slag system, the local corrosion occurs through a mechanism similar to that of the "tears of wine."[75] It seems that a trivial, everyday life phenomenon, observed by Thomson,[75] actually held the key to solve problems that are significant in engineering after more than a hundred years.

It has already been apparent that the mechanism of the wet corrosion of the metal can be explained well based on the concept of a local battery. As described in Section 4.2.2, the mechanism of local corrosion of the oxide refractory is being scientifically systematized based on interfacial physical chemistry processes such as the Marangoni effect.

In practice, these achievements will surely move the starting point of the development and improvement of the refractory materials forward and contribute to the development of high-temperature interfacial physical chemistry as a discipline. From a practical viewpoint, it seems to be one issue to find more effective measures to prevent the local corrosion in the future through quantitatively describing the rate of local corrosion caused by the Marangoni effect with the condition of real operation in mind.

4.2.3 MOTION OF FINE PARTICLES IN LIQUID UNDER INTERFACIAL TENSION GRADIENTS

When fine particles such as non-metallic inclusions and bubbles dispersed in molten steel are entrapped in steel at the time of solidification, their states of being in the steel closely affect the material properties of the steel. It is becoming clear that the interfacial properties have a significant influence on the behavior of fine particles in molten steel.

Specifically, fine bubbles and solid particles move from a region of high interfacial tension to that of low interfacial tension under the interfacial tension gradient, which in turn can be a main factor dominating the behavior of fine particles in each process of refining where the concentration gradient of the interfacially active components exists, e.g., molten steel close to the inner wall of the immersion nozzle for continuous casting or at the front side of the solidifying interface.

The driving force of particle motion caused by the pressure difference is expressed by the flowing Laplace's equation (4.2):

$$p^p - p^l = \frac{2\gamma^{pl}}{r_p},$$ (4.2)

where p^p and p^l are the pressure in the particle and liquid phases, respectively, γ^{pl} is the interfacial tension between particle and liquid phases, and r_p is the radius of the particle (sphere).

Kaptay et al.[76] named this force as F_{ML} after Mukai et al.[77] However, this force is rather considered as the effect based on the Laplace's equation (4.2). Therefore, in the naming of this force, the terminology "Mukai–Lin–Laplace (MLL) effect" would be more appropriate.

This phenomenon is different from the Marangoni convection characterized by the liquid movement and takes the form of particle movement instead. Notwithstanding, the phenomenon can be termed as the Marangoni effect in a broad sense in that it is caused by the interfacial tension gradient.

4.2.3.1 Motion of Fine Bubbles in Aqueous Solution under Surface Tension Gradient[77] †

In the case of a surface tension gradient based on the concentration gradient of the surface-active component $C_{18}H_{29}SO_3Na$ along the surface of fine hydrogen bubbles in an aqueous solution, the bubbles move from a region of high surface tension to that of low surface tension. This phenomenon was confirmed by direct observation. In addition, the rate of movement (terminal velocity) of the particles was derived using Laplace's equation (4.2) under the condition where the fine bubbles can be regarded as a rigid particle (sphere), and the following equation was obtained:‡

$$V_I = \frac{-4rK_\gamma}{9\eta},$$ (4.3)

(Continued on page 107)

---

‡ Translation supervisor note:
  Laplace's equation:

$$p^p - p^l = \frac{2\gamma^{pl}(x)}{r_p}$$

where p^p is the pressure of the particle and p^l is that of the liquid phase, $\gamma^{pl}(x)$ is the surface/interfacial tension between the particle and the liquid phase at x, and r_p is the radius of the particle. The infinite small area of the ring (see figure), ds, is

$$ds = 2\pi r_p \cos\theta \frac{dx}{\cos\theta} = 2\pi r_p dx$$

(Note that the width of the ring is not dx but $dx/\cos\theta$.)

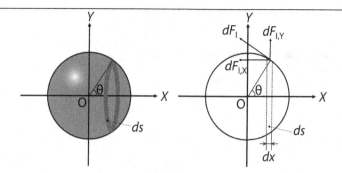

Therefore, the force induced on this infinite small area owing to the surface tension gradient is:

$$dF_I = -\Delta p \cdot ds = -\left(p^p - p^l\right) \cdot 2\pi r_p dx = -\frac{2\gamma^{pl}(x)}{r_p} \cdot 2\pi r_p dx = -4\gamma^{pl}(x)\pi dx$$

The force acting in the direction of the X-axis, $dF_{I,X}$ is:

$$dF_{I,X} = \cos\theta \cdot dF_I = -\frac{x}{r_p} \cdot 4\gamma^{pl}(x)\pi dx$$

Hence:

$$F_{I,X} = -\frac{4\pi}{r_p} \cdot \int_{-r_p}^{r_p} x \cdot \gamma^{pl}(x)dx$$

The net force acting in the direction of the Y-axis, $F_{I,Y}$, is zero.
Therefore:

$$F_I = F_{I,X} = -\frac{4\pi}{r_p} \cdot \int_{-r_p}^{r_p} x \cdot \gamma^{pl}(x)dx$$

When the surface tension gradient is constant, i.e., $d\gamma^{pl}(x)/dx = K$:
$$\gamma^{pl}(x) = Kx + C$$

Thus:

$$F_I = -\frac{4\pi}{r_p} \cdot \int_{-r_p}^{r_p} x \cdot \gamma^{pl}(x)dx$$

$$= -\frac{4\pi}{r_p} \cdot \int_{-r_p}^{r_p} x \cdot \left(Kx + C\right)dx$$

$$= -\frac{8}{3}\pi r_p^2 K$$

Two forces, F_I and F_D, act on the particle.
As derived above, the force due to the surface/interfacial tension gradient, F_I, can be expressed as:

$$F_I = -\frac{8}{3}\pi r_p^2 K$$

where r_p is the radius of the particle and K is the surface tension gradient, $d\gamma^{pl}(x)/dx$.

The force acting on the particle moving through a viscous fluid, F_D, is:

$$F_D = 6\pi\eta r_p V_I$$

where η is the viscosity and V_I is the velocity of the particle.
(cf. Stokes' law)
Thus, the net force acting on the particle is:

$$F = -\frac{8}{3}\pi r_p^2 K - 6\pi\eta r_p$$

This force is equal to:

$$F = ma = \frac{4}{3}\pi r_p^3 \rho_P \frac{dV_I}{dt}$$

where ρ_P is the density of the particle. Hence:

$$\frac{dV_I}{dt} = \frac{-\dfrac{8\pi}{3}r_p^2 K - 6\pi\eta r_p V_I}{\dfrac{4}{3}\pi r_p^3 \rho_P}$$

By solving the differential equation, we obtain:

$$V_I = \frac{1}{6\pi\eta r_p} \cdot \frac{8\pi}{3}r_p^2 K \left\{ \exp\left(\frac{-6\pi\eta r_p t}{\dfrac{4}{3}\pi r_p^3 \rho_P}\right) - 1 \right\}$$

Here:

$$\exp\left(\frac{-6\pi\eta r_p t}{\dfrac{4}{3}\pi r_p^3 \rho_P}\right) \approx 0$$

Therefore:

$$V_I = \frac{-4 r_p K}{9\eta}$$

(End of the derivation of Equation (4.3))
When the particle is located close to the wall, a correction factor β must be added for F_D as follows.

$$F_D = 6\pi\eta r_p V_I \beta$$

In addition, β is expressed as:

$$\beta = 1 + \frac{9 r_p}{8H}$$

where H is the distance between the particle and the wall. In this case, V_I becomes:

$$V_I = \frac{-4 r_p K}{9\eta}\left(1 + \frac{9 r_p}{8H}\right)^{-1}$$

where V_I is the velocity of the particle, K_γ is the surface tension gradient in the direction of motion, and η is the viscosity of the liquid phase.

Using Equation (4.3), the rate of movement of the bubbles with a diameter of 100 μm or less (experimental result) can be reasonably described. Under the condition where the vertical surface tension gradient is 0.157 N/m² (surface tension is higher in the upper region), a strange phenomenon in which the bubbles overcome buoyancy and sink is clearly observed when their diameters become less than 70 μm.

From the viewpoint of fluid mechanics, it can be considered that the above-mentioned movement of fine bubbles occurs not only for bubbles but also for droplets and solid particles. Therefore, there is a high possibility that this phenomenon is closely related to each elementary process in refining involving bubbles and inclusions.

4.2.3.2   Engulfment and Pushing of Fine Particles at the Solidification Interface[†]

It is interesting to know whether the fine particles in the liquid phase are trapped in or extruded from the solid–liquid interface during the solidification. This phenomenon is important not only for steel refining but also for materials engineering in general. For details on the research on this topic, please refer to my review article.[78] As is well known, the expelled solute generally accumulates on the liquid phase side, where the solidification occurs; as the solidification progresses, the concentration gradient of the solute is formed (Figure 4.22). Therefore, if the solute is an interfacially active component, particles in the region where the concentration gradient is present should be pulled to the solid–liquid interface and easily trapped there by force $\mathbb{F}_I$, which is caused by the interfacial tension gradient based on this concentration gradient (Figure 4.22). However, as pointed out in the above-mentioned review,[78] the effect of this concentration gradient has not been considered in the previous studies.

To clarify this phenomenon, the behavior of the fine bubbles on the front side of the solidified interface in an aqueous solution was directly observed.[4] The results found that hydrogen bubbles on the front side of the vertical solidified interface of the aqueous solution containing a small amount of surfactant $C_8H_{17}SO_3Na$ suddenly changed its trajectory, when they approached the interface at about 100 μm, from vertical to horizontal movement toward the solidified interface rapidly, and were trapped. Because of these processes, the so-called "dirty ice," whose inside is full of bubbles, was formed (Figure 4.23). However, when NaCl, which hardly changes the surface tension of water, was added at the same concentration as that of the above surfactant to an aqueous solution, the hydrogen bubbles only rose vertically and were not trapped, resulting in a "clean ice" containing no bubbles (Figure 4.24). Furthermore, under the microgravity

---

When the surface tension gradient, K, is not constant and the particle is located far from the wall:

$$V_I = -\frac{2}{3\eta r_p^2} \cdot \int_{-r_p}^{r_p} x \cdot \gamma^{pl}(x)\,dx$$

When the surface tension gradient, K, is not constant and the particle is located close to the wall:

$$V_I = -\frac{2}{3\eta r_p^2} \cdot \left(1 + \frac{9r_p}{8H}\right)^{-1} \cdot \int_{-r_p}^{r_p} x \cdot \gamma^{pl}(x)\,dx$$

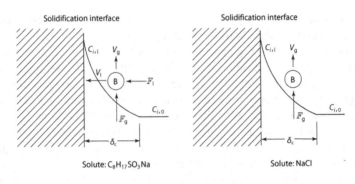

B: bubble
$C_{i,i}$: concentration of the solute near interface
$C_{i,0}$: concentration of the solute in bulk phase
F_i : force caused by the surface tension gradient
V_i: velocity originating from F_i
F_g: buoyancy (caused by gravity)
V_g: velocity originating from buoyancy
δ_c : thickness of the concentration boundary layer

FIGURE 4.22   Forces acting on a bubble and the velocity of the bubble near the solidifying interface (solidifying from left to right) of the aqueous solution. (Adapted from reference 3.)

condition obtained by the parabolic flight using jet aircraft,[‡] it was clearly confirmed by the experiments using a Mach–Zehnder interferometer that the fine bubbles moved toward the interface only in the region where the concentration gradient on the front side of the solidification interface was detected.[79] The behavior of the hydrogen bubbles described above can be basically expressed by an equation that combines Equation (4.3) with the concentration distribution of the surface-active component in the aqueous solution at the solidification interface.[4] An analysis of the engulfment and pushing of bubbles and inclusions at the solidification interface of molten iron, which took the above-mentioned concentration gradient effect into consideration, has already pointed out that the bubbles and inclusions might be rapidly pulled to the solidification interface in the presence of, e.g., O, S, and Ti, as a solute component.[80] The above-mentioned experiments and analysis of the water-model system further proved this possibility. Furthermore, the relation between residual bubbles and inclusions in steel and the concentration of interfacially active components such as O, S, N, Ti, B, and Nb in a real operation could be formulated using Equation (4.3) and reasonably described.[81,82,*] This result can be a guideline for obtaining steel with few bubbles and inclusions.

[‡] Translation supervisor note: Besides the parabolic flight, there are several ways to obtain the microgravity condition, e.g., sounding rocket, drop shaft, and International Space Station (ISS). Utilizing the microgravity condition for the experiments is a promising way to understand the phenomena, precise measurements of material properties, etc. For example, a project to measure the interfacial tension between the molten steel and slag by the levitation technique at ISS is ongoing under the management of the Japan Aerospace Exploration Agency (JAXA).

[*] Translation supervisor note: After these two papers, the author published the following paper: T. Matsushita, K. Mukai and M. Zeze, Correspondence between Surface Tension Estimated by a Surface Thermodynamic Model and Number of Bubbles in the Vicinity of the Surface of Steel Products in Continuous Casting Process, *ISIJ Int.*, 53 (2013), 18.

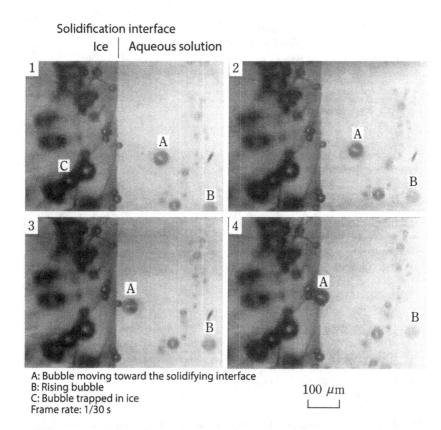

Solidification interface

Ice | Aqueous solution

A: Bubble moving toward the solidifying interface
B: Rising bubble
C: Bubble trapped in ice
Frame rate: 1/30 s

100 μm

FIGURE 4.23 Behavior of the bubbles in front of the solidifying interface of an aqueous solution of $C_8H_{17}SO_3Na$. (Adapted from reference 3.)

4.2.3.3 Clogging of the Immersion Nozzle[83,84]

In the continuous casting of molten steel, the inclusions (mainly Al_2O_3) deposit and accumulate on the inner wall of the immersion nozzle, causing a nozzle clogging, which has become a serious problem in the operation of continuous casting. The motion driven by the interfacial tension gradient might also be closely related to this clogging phenomenon. At the interface between the currently widely used SiO_2-containing Al_2O_3-graphite nozzle material and molten steel, the concentration gradients of $\underline{Si}$ and $\underline{C}$ are formed at the interface in the process in which SiO and CO gases produced by the reaction, shown in Equation (4.4), in the nozzle dissolve in the molten steel.

$$SiO_2\,(s) + C\,(s) = SiO\,(g) + CO\,(g). \qquad (4.4)$$

Since both components reduce the Al_2O_3–molten steel interfacial tension, the inclusions are pulled to the interface (see Figure 4.25). If the process in which the inclusions in molten steel approaching near the interface move to the nozzle wall is playing a dominant role in the process of deposition of the inclusions, the

Solidification interface
Ice | Aqueous solution

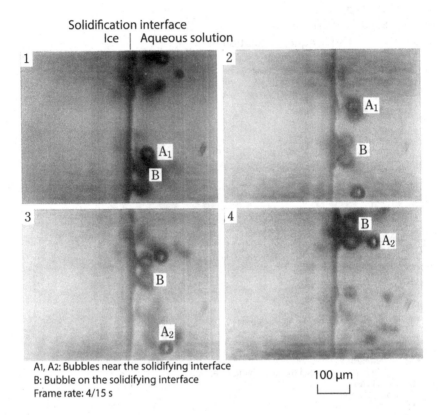

A1, A2: Bubbles near the solidifying interface
B: Bubble on the solidifying interface
Frame rate: 4/15 s

100 μm

FIGURE 4.24 Behavior of the bubbles in front of the solidifying interface of the NaCl aqueous solution. (Adapted from reference 3.)

deposition of the inclusions should be suppressed if the nozzle material that does not produce SiO and CO gases or does not supply such gas to the molten steel is used. Then, to interrupt the transport path of SiO and CO gases generated in the nozzle material to molten steel and to suppress the gradient of the interfacial tension between inclusions and molten steel caused by the concentration gradient of $\underline{Si}$ and $\underline{C}$ around the inner surface of the nozzle, a high-purity alumina layer was attached on the inner surface of the nozzle. As a result, the deposition of the inclusions was considerably suppressed, and the nozzle clogging was suppressed. The obtained result in the real operation was satisfactory. In continuous casting operations, it is desirable to reduce the amount of argon gas injection as much as possible. In this respect, the use of high-purity Al_2O_3 is noteworthy in that the deposition of inclusions is extremely suppressed even when the amount of the argon gas injection is reduced.[81]

Furthermore, the development of the following nozzle material may be possible by more actively applying the above idea. One example is the material that provides a situation where the interfacial tension becomes low as the inclusions (mainly Al_2O_3) move away from the inner wall of the nozzle, that is the interfacial tension gradient in the reverse direction of that shown in Figure 4.25. The author has proposed the

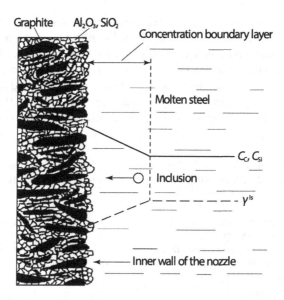

FIGURE 4.25 Behavior of the inclusions in molten steel near the interface between molten steel and the inner wall of the immersion nozzle. (Adapted from reference 81.)

use of dolomite instead of Al_2O_3 as the oxide component of the nozzle material. Dolomite has a high desulfurization ability and absorbs sulfur in molten steel, $\underline{S}$. Therefore, it can be used as a desulfurization material. If the desulfurization reaction rate is dominant, the region with the concentration gradient in which the $\underline{S}$ concentration increases with the distance from the inner wall will be formed. In this case, the inclusions in this region are subjected to force F_I in the direction away from the inner wall due to the interfacial tension gradient, and it is possible to more actively prevent the deposition of the inclusions. Recently, a dolomite-graphite nozzle material that can withstand the practical use was developed,[85] and very good results were obtained in the actual operation.

In Section 4.2.3, solid particles have been treated as objects which can be moved under interfacial tension gradient, similar to bubbles. In the experiment on the ground, it has already been confirmed that the sedimentation rate of fine graphite particles becomes low in pure water under temperature gradient (temperature is higher in the upper region),[86,87] and leukocytes in swine blood move to the high-concentration region of chemokine under the chemokine concentration gradient.[88] However, the movement of fine solid particles due to the interfacial tension gradient under microgravity has not yet been confirmed. Furthermore, a method that can directly confirm the movement of the inclusions in molten steel under the interfacial tension gradient has not yet been found.

In Section 4.2.3.3, the motion driven by the interfacial tension gradient was treated to be applicable to the inclusions in molten steel. As a result, we achieved concrete results in the clarification and countermeasures for the important technical issues in the steel refining process, as described above. This is why this topic was daringly selected.

However, an essential clarification regarding the movement of the solid fine particles described above still awaits future work. We sincerely wish for progress in the research in this field. From the viewpoint of academic study, the clarification of such fine particle behavior under the interfacial tension gradient will create and expand the relation between the chemical potential at non-equilibrium and hydrodynamics, i.e., a new field of physicochemical hydrodynamics. Furthermore, this phenomenon is not limited to steel refining but is likely an important phenomenon in a wide range of natural sciences.[‡]

1. Translation supervisor note: The video clip is available for free at https://www.crcpress.com/Interfacial-Physical-Chemistry-of-High-Temperature-Melts/Mukai-Matsushita/p/book/9780367210328

Contents: Interfacial phenomena of high-temperature melts and materials processing – In situ observation – (00:00)
1. Behavior of the injected argon gas in the nozzle and mold in the continuous casting of molten steel (water model experiment) (00:05)
1-1 Behavior in the nozzle (00:14)
1-2 Behavior in the mold (02:13)
2. Penetration behavior of molten slag and metal into refractories (02:48)
2-1 Penetration of molten slag (02:53)
2-2 Penetration of molten metal (03:51)
3. Marangoni effect in high-temperature melt (06:50)
3-1 Marangoni convection in liquid column caused by the temperature gradient (07:17)
3-2 Expansion and contraction of the slag droplets caused by electric potential change (10:30)
4. Local corrosion at the slag surface and slag–metal interface of refractory (12:00)
4-1 System in which the surface (interfacial) tension increases with the dissolution of the refractory component (movement of the slag film based on the concentration gradient) (12:19)
4-1-1 Local corrosion of the solid silica (SiO_2) at the lead oxide–silica ($PbO–SiO_2$) slag surface (12:31)
4-1-2 Local corrosion of solid silica (SiO_2) at the interface between the $PbO–SiO_2$ slag and Pb (15:47)
4-2 System in which the surface tension decreases with the dissolution of refractory component (18:38)
4-2-1 Local corrosion of solid silica at the surface of the iron oxide–silica ($FeO–SiO_2$) slag (19:00)
4-3 Local corrosion of oxide–non-oxide composite refractory at the slag/molten–iron interface (20:50)
4-3-1 Local corrosion of alumina–graphite refractory (21:13)
4-3-2 Local corrosion of magnesia–carbon refractory (23:39)
5. Motion of fine bubbles in aqueous solution under the surface tension gradient (24:41)

[‡] Translation supervisor note: In addition to Ref. 88, the author has published the following studies in the field of bioengineering.S. Harada, M. Tamagawa, K. Mukai, Y. Furukawa, and I. Yamanoi, "Observation of Neutrophil Motion by Gradient of Cytokine Concentration in Water," Proceedings of the 18th Bioengineering Conference, Niigata, Japan (2006) p. 135–136. M. Tamagawa, Y. Kubomoto, K. Mukai, and Y. Furukawa, "Observation and Analysis of neutrophile motion by concentration gradient," Proceedings of the 56th National Congress of Theoretical and Applied Mechanics (NCTAM), Tokyo, Japan (2007) p. 291.

5-1 Motion in a state where the surface tension gradient is in a steady state (24:55)
5-2 Motion near the solidification interface of aqueous solution (26:36)

REFERENCES

1. H. Sakao and K. Mukai: *Tetsu-to-Hagané*, **63** (1977), 513.
2. Z. Wang, K. Mukai, and D. Izu: *ISIJ Int.*, **39** (1999), 154.
3. Z. Wang, K. Mukai, Z. Ma, M. Nishi, H. Tsukamoto, and F. Shi: *ISIJ Int.*, **39** (1999), 795.
4. Z. Wang, K. Mukai, and I. J. Lee: *ISIJ Int.*, **39** (1999), 553.
5. K. Mukai, Z. Tao, K. Goto, Z. Li and T. Takashima: *Taikabutsu*, **53** (2001), 390.
6. T. Matsushita, K. Mukai, T. Ohuchi, I. Sasaka, and J. Yoshitomi: *Taikabutsu*, **55** (2003), 120.
7. T. Matsushita and K. Mukai: *Tetsu-to-Hagané*, **90** (2004), 429.
8. M. Ohkubo, Y. Miyashita and R. Imai: *Tetsu-to-Hagané*, **54** (1968), S59.
9. H. Suito, H. Inoue, and R. Inoue: *ISIJ Int.*, **31** (1991), 1381.
10. K. Wasai and K. Mukai: *ISIJ Int.*, **42** (2002), 467.
11. H. Shingu and K. Ishihara: *Bull. Japan Inst. Metals Mater. Japan*, **25** (1986), 16.
12. K. Wasai and K. Mukai: *Metall. Mater. Trans. B*, **30B** (1999), 1065.
13. K. Wasai, K. Mukai, and A. Miyanaga: *ISIJ Int.*, **42** (2002), 459.
14. Y. Ogawa and N. Tokumitsu: 6th Int. Iron Steel Cong., Vol. 1, ISIJ, Tokyo (1990), 147.
15. K. Mukai: *Tetsu-to-Hagané*, **77** (1991), 856.
16. K. Mukai, T. Nakamura, and H. Terashima: *Tetsu-to-Hagané*, **78** (1992), 1682.
17. Y. Zhang and R. J. Fruehan: *Metall. Mater. Trans. B*, **26B** (1995), 803.
18. M. Byrne and G. R. Belton: *Metall. Trans. B*, **14B** (1983), 441.
19. J. Lee and K. Morita: *ISIJ Int.*, **43** (2003), 14.
20. G. S. Ershov and V. M. Bychev: *Russ. Metall.*, **4** (1975), 45.
21. Z. Jun and K. Mukai: *ISIJ Int.*, **38** (1998), 1039.
22. N. Hirashima, R. T. C. Choo, J. M. Toguri, and K. Mukai: *Metall. Mater. Trans. B*, **26B** (1995), 971.
23. Z. Jun and K. Mukai: *ISIJ Int.*, **38** (1998), 220.
24. Z. Jun and K. Mukai: *ISIJ Int.*, **39** (1999), 219.
25. Z. Jun, F. Shi, K. Mukai, and H. Tsukamoto: *ISIJ Int.*, **39** (1999), 409.
26. T. Nakamura, K. Yokoyama, F. Noguchi, and K. Mukai: Molten Salt Chemistry and Technology, Trans. Tech. Publications, Switzerland, *Mater. Sci. Forum*, **73–75** (1991), 153.
27. K. Morinaga, T. Yanagase, Y. Ohta, and Y. Ueda: *TMS Paper Selection* (1979), A-79–18.
28. N. Imaishi, S. Yasuhiro, T. Nakamura, and K. Mukai: 18th Int. Symp. Space Technol. Sci., Kagoshima, Japan (1992), 2179.
29. K. Ishizaki, N. Araki and H. Murai: *J. Jpn. Weld. Soc.*, **34** (1965), 146.
30. C. R. Heiple and J. R. Roper: *Welding J.*, **61** (1982), 97s.
31. C. R. Heiple, J. R. Roper, R. T. Stagner, and R. J. Aden: *Welding J.*, **62** (1983), 72s.
32. K. Mukai, M. Watanabe, T. Yamada, N. Takiuchi and S. Shinozaki: *J. Jpn. Inst. Met.*, **55** (1991), 36.
33. N. Takiuchi, T. Tani, Y. Tanaka, N. Shinozaki, and K. Mukai, *J. Jpn. Inst. Met.*, **55** (1991), 180.
34. S. I. Popel, B. V. Tsarevski, V. V. Pavlov, and E. L. Furman: *Izv. Akad. Nauk SSSR Met.* (1975), 54.
35. Z. Niu, K. Mukai, Y. Shiraishi, T. Hibiya, K. Kakimoto, and M. Koyama: *J. Japan. Assoc. Cryst. Growth*, **24** (1997), 369.
36. K. Mukai, Z. Yuan, K. Nogi, and T. Hibiya: *ISIJ Int.*, **40** (2000), Supplement S148.

37. T. Hibiya, S. Nakamura, T. Azami, M. Sumiji, N. Imaishi, K. Mukai, K. Onuma, and S. Yoda: *Acta Astronaut.*, **48** (2001), 71.
38. T. Hibiya: *J. Mater. Sci.*, **40** (2005), 2417.
39. B. V. Patrov: *Surface Phenomena in Matallurgical Processes*, ed. A. I. Belyaev, Consultants Bureau Enterprises, Inc. (1965), 129.
40. K. Mukai, J. M., Toguri, I. Kodama, and J. Yoshitomi: *Can. Metall. Q.*, **25** (1986), 225.
41. R. T. C. Choo and J. M. Toguri: *Can. Metall. Q.*, **31** (1992), 113.
42. K. Mukai: *Mater. Jpn.*, **34** (1995), 395.
43. K. Mukai, J. Yoshitomi, T. Harada, K. Hurumi and S. Fujimoto: *Tetsu-to-Haganè*, **70** (1984), 541.
44. R. Brückner: *Glastech. Ber.*, **53** (1980), 77.
45. R. Brückner: *Glastech. Ber.*, **40** (1967), 451.
46. M. Dunkel and R. Brückner: *Glastech. Ber.*, **53** (1980), 321.
47. T. S. Busby: *Glass Technol.*, **20** (1979), 117.
48. H. Jebsen-Marwedel: *Glastech. Ber.*, **29** (1956), 233.
49. E. Vago and C. E. Smith: VII Int. Congress Glass, Brussels, **II** (1965), 1.2/62.1.
50. J. Löffler: *Glastech. Ber.* (1968), 513.
51. W. F. Caley., B. R. Marple and C. R. Masson: *Can. Metall. Q.*, **20** (1981), 215.
52. A. Sendt: VIIe Congres Int. du Verre, Bruxelles (1965), 352.
53. K. Schulte: *Glastech. Ber.*, **50** (1977), 181.
54. Y. Iguchi, G. J. Yurek, and J. F. Elliott: III Int. Iron Steel Congress, ASM, Chicago (1978), 346.
55. K. Mukai, A. Iwata, T. Harada, J. Yoshitomi and S. Fujimoto: *J. Jpn. Inst. Met.*, **47** (1983), 397.
56. K. Mukai, T. Harada, T. Nakano, K. Hiragushi: *J. Jpn. Inst. Met.*, **49** (1985), 1073.
57. K. Mukai, T. Harada, T. Nakano, K. Hiragushi: *J. Jpn. Inst. Met.*, **50** (1986), 63.
58. M. Hino, T. Ejima and M. Kameda: *J. Jpn. Inst. Met.*, **31** (1967), 113.
59. K. Mukai, T. Masuda, K. Gouda, T. Harada, J. Yoshitomi and S. Fujimoto: *J. Jpn. Inst. Met.*, **48** (1984), 726.
60. K. Mukai, K. Gouda, J. Yoshitomi, and K. Hiragushi: III Int. Conf. Molten Slags Fluxes, The Inst. of Met., London, (1988), 215.
61. Z. Tao, K. Mukai, S. Yoshinaga, and M. I. Ogata: *Taikabutsu*, **50** (1998), 316.
62. J. Yoshitomi, T. Harada, K. Hiragushi, and K. Mukai: *Tetsu-to-Haganè*, **72** (1986), 411.
63. J. Yoshitomi, K. Hiragushi, and K. Mukai: *Tetsu-to-Haganè*, **73** (1987), 1535.
64. Z. Yu and K. Mukai: *J. Jpn. Inst. Met.*, **59** (1995), 806.
65. Z. Yu and K. Mukai: *J. Jpn. Inst. Met.*, **56** (1992), 1137.
66. K. Mukai, T. Masuda, J. Yoshitomi, T. Harada and S. Fujimoto: *Tetsu-to-Haganè*, **70** (1984), 823.
67. Z. Tao, K. Mukai, and M. Ogata: *Taikabutsu*, **50** (1998), 460.
68. T. B. King: *Trans. Met. Soc. AIME*, **230** (1964), 1671.
69. Z. Yu: PhD thesis, Kyushu Institute of Technology, Japan (1993).
70. F. Hauck and J. Pötschke: *Arch. Eisenhüttenw.*, **53** (1982), 133.
71. K. Mukai, J. M. Toguri, and J. Yoshitomi: *Can. Metall. Q.*, **25** (1986), 265.
72. K. Mukai, J. M. Toguri, N. M. Stubina, and J. Yoshitomi: *ISIJ Int.*, **29** (1989), 469.
73. T. Kii, K. Hiragushi, H. Yasui and K. Mukai: Unified International Technical Conference on Refractory, 4th Biennial Worldwide Conference on Refractories, TARJ, Kyoto, **3** (1995), 379.
74. Z. Li, K. Mukai, and Z. Tao: *ISIJ Int.*, **40** (2000), Supplement, S101.
75. J. Thomson: *Philos. Mag., Ser.*, **4**, 10 (1855), 330.
76. G. Kaptay and K. Kelemen: *ISIJ Int.*, **41** (2001), 305.
77. K. Mukai and W. Lin: *Tetsu-to-Haganè*, **80** (1994), 527.
78. K. Mukai: *Tetsu-to-Haganè*, **82** (1996), 8.

79. K. Mukai, Y. Furukawa, H. Segawa and E. Yokoyama: Performed at Komaki Airport, Nagoya, Japan in December 2001 (not published).
80. K. Mukai and W. Lin: *Tetsu-to-Hagané*, **80** (1994), 533.
81. K. Mukai and M. Zeze: *Steel Res.*, **74** (2003), 131.
82. K. Mukai, L. Zhong, and M. Zeze: *ISIJ Int.*, **46** (2006), 1810.
83. K. Mukai, R. Tsujino, I. Sawada, M. Zeze and S. Mizoguchi: *Tetsu-to-Hagané*, **85** (1999), 307.
84. R. Tsujino, K. Mukai, W. Yamada, M. Zeze, and S. Mizoguchi: *Tetsu-to-Hagané*, **85** (1999), 362.
85. K. Ogata, J. Amano, K. Morikawa, J. Yoshitomi and K. Asano: Krosak Harima Technical Report, **152** (2004), 24.
86. H. Segawa: MSc thesis, Kyushu Institute of Technology (2003).
87. Hi. Nakaooji: MSc thesis, Kyushu Institute of Technology (2004).
88. M. Tamagawa, K. Mukai, and Y. Furukawa: Particulate Process in the Pharmaceutical Industry, Montreal, Canada (2005), 10.

Index

Printed in the United States
by Baker & Taylor Publisher Services